BHAGVAT GITA

DIVINE SONG ON CREATOR, CREATION AND CREATURE
STANZA WISE PERCEPTION OF HIDDEN SCIENCE

Bharatiya have habit of boasting that all modern scientific discoveries
are in Vedik literature. It can be added that not only the scientific
discoveries exist, but the Sanskrit speaking civilization was much
more advanced scientifically than modern civilization. Technology
for Growing humans outside womb in an artificial devise was known
to scientists of Mahabharat era.

First ever book on hidden science in GITA stanzas
STANZAVISE ANALYTICAL STUDY
SHRIMAD BHAGWAT GITA

DR. B.G. MATAPURKAR

SONG OF SCIENCE

SHRIMAD BHAGWAT GITA

SCIENCE IN GITA

A SCIENTIFIC SONG ON CREATOR,
CREATION AND CREATURE

STANZAVISE ANALYTICAL STUDY
A SCIENTIFIC SCRUTINY

DR. B.G. MATAPURKAR

INDIA · SINGAPORE · MALAYSIA

Notion Press

No. 8, 3rd Cross Street,
CIT Colony, Mylapore,
Chennai, Tamil Nadu – 600 004

First Published by Notion Press 2020
Copyright © Dr. B.G. Matapurkar 2020
All Rights Reserved.

ISBN 978-1-63714-582-1

Lord Krishna convinces Arjun on the battlefield of Kurukshetra. During such conviction many scientific facts are mentioned as examples. Attempt is made to explain such hidden scientific facts using original work found in various relevant, available Sanskrit literature on the subjects.

DEDICATED TO THE ARCHAEOLOGISTS OF THE WORLD
FOR THEIR ZIEST TO EXPLORE THE FACTS AND TRUTH
OF PAST CIVILIZATIONS

AND

YONGER GENERATIONS
DESCENDENTS OF
SANSKRIT SPEAKING CIVILIZATION OF GLOBE

CONTENTS

ABSTRACT

Gita is divine scientific song on creator, creation, and creature,. It deals with scientific study of nature and supernatural elements for ultimate human benefits. As per the Sanskrit literature the manifest material universe is "LINGATMAK and MAITHUNIK" oriented. It means it is based on union and interaction of two substances (DRAVYA-द्रव्य) for newer manifestations. Hence the creator is called "SHIV LING". Science of VED reveals that there is pre embryonic creation in the form of RAYI (material manifestation) and PRAN (spiritual energy) from BRAHM. Union of male and female for newer generation and Oxygen with hydrogen for new substance water formation. Water with heat for steam formation, water with cold temperature for ice formation. Sulphur with Oxygen for sulphur di oxide formation and so on. Everything in manifest universe is living. Every material has life force or energy the CHETAN TATVA in it. To understand this mystic fact MEDITATION is essential. **Meditation is a silent dynamism. Brain understands more in silence; hence yield is more with less energy consumption. Supernatural "Brain faculties" get opened up in dynamic silence.**

All manifest material world has birth, growth, change, life span and death. Difference is with "TIME CYCLE". Galaxies, Stars, planets, star constellations – "nakshtra", all are living. All have birth, growth, change, multiplication and death. On earth, time cycle is small compared to BRAHM, time cycle of Brahm is small compared to Vishnu and so is Mahesh (SHIV). Even SHIV has life span. There is different supreme authority behind the whole CELESTIAL DRAMA in space (Aakash).

There are ten different PRAN each having different function and each flow through different flow channels. Analytical presentation in the perspective of modern science is discussed in the book.

Author's opinion

PRAYER OR SHANTI MANTRA
IN SANSKRIT LITERATURE

VED (means knowledge) have developed SHANTI or prayer mantra. These are scientifically oriented. Every prayer is scientific and full of knowledge. It is hardly having any religious (modern religion, sect or cult or communalism) intentions. Every Ved or Upnishad to start with, have their prayer shanti mantra. Some of the popular prayer mantra are given below with scientific zest in them:

Om purnamadah purnmidam purnat purn mudachyate I
punrsya purnmaday purn mevav shishyate II
Om shanti! Shanti!! Shanti!!!

(*Brihdaranyok upnishad (Shankar bhashya). Code 577, Gita Press Gorakhpur. ISBN 81-293-0247-0, page 27.*)

Purnamadah (पूर्णमदः) means "That" (parabrahma) is complete, Purnmidam (पूर्णमिदम्) "this" (working brahma) is complete. Because complete can give rise to complete and exist together. If complete is taken out of complete still complete remains. At the time of annihilation

(PRALAY KAAL), working brahma and parabrahma absorbing in each other, recreate complete brahm again.

What is "that" and "this"? If one understands BRAHM, it has two words in it. BRAHA (ब्रह) and AHAM (अहम्). Brah means out or away and Aham means me or near. Therefore, the words used are for the BRAHMA only. That means universal brahm and this means brahm entrapped in material body. Scientific meaning can be explained by use of stem cell generation (Author's research).

Stem cell is complete. After fertilization of ovum with sperm, the embryo growth depends upon multiplication of cell formed on fertilization. Single cell fertilized ovum, forms similar cells by division. These cells are complete by them self, i.e. it needs no help what so ever. It is auto nourished by itself. These can generate their own protein themselves. Cells divide and re-divide to form foetus. New cells are formed which are complete and remaining cells are also complete. New cells are capable of, producing complete cells again. (Author considers success of his research is due to this automation of complete cells or BRAHM).

Other scientific explanation can be cited is single celled AMOEBA. It is complete and can divide and form another complete amoeba. At the same time the original amoeba is remaining complete living being. This scientific truth is with all single celled living beings.

PRASHNOPNISHAD, Mundkopnishad, Shanti Mantra:

Om bhadram karnebhi shrunuyam devah I
bhadram pashyemakshbhiryajatrah I
sthire angeh trishtvagam sastanubhih I
vyashem devhitamyadayuh II
swasti na indro vridhashravah swastinah pusha vishvavedah I
swasti nastarksho arishtnemih swastino brihaspatirdadhatu II
Om shanti! Shanti!! Shanti!!!

Oh! Super soul let us listen pious and auspicious, let us see pious and auspicious, with normal and healthy body and all appendages, while praying and praising your virtues let us live the life bestowed and allotted by nature. May ancient and famous "Rain God", the knower of everything the Sun god-PUSHNA, the lord of swift motion, celestial bird who protect us from harm the wind, grant us what is auspicious and pious. The god of intellect, spirituality and wisdom Brihaspati bless us. OHM peace!, peace!!, peace!!! be there.

Rig ved mantra: **(RIGVED – mandala 3. Sutra 62. Mantra 10). Composer of Gayatri Mantra is Sage VISHWAMITRA.**

Om bhu rbhuvah swah I
tatsviturvarenyam bhargo devasya dhimahi I
dhiyo yonah prachodayatI

Prayer to absolute truth (parambrahma), creator of all universe, protector of universe, to stimulate intellect and bestow true knowledge upon us.

There is not the least tinge of modern religion in any form in this prayer. The prayer is only for knowledge and awakening intellect.

BHU: Life giver to material world, Presiding deity of earth-prithvi lok The PRAN.

BHUVA: Protector and nourisher of living beings, Presiding deity of ANTARIKSHA lok, the APAN.

SWAH: Extensively and comprehensively involved in universe. Deity of DYO lok, the VYAN.

Generator of whole universe, giver of knowledge, worship able, pure form on which we meditate. Prayer to force us in right direction. The Gayatri Mantra is to pray for directing intellect and mind-set of all.

Yajurved shanti mantra:

Om dyoh shantir antriksha shantihi prithvi
shantirapah shantir aushadhaya shanti I
Vanaspatayah shantirvishve devah shantirbrahm shanti,
Sarvam shanti Shantirev shanti, Sa Ma shantiredhi II
Om shanti shanti shanti

Om! the god of universe grant peace in all the three abodes, the space, water and earth (Jal, Thal, Nabh), Antariksh, fire, wind, medicines, vegetation of forest, the whole universe, in hearts of all living beings, in me and in all, in every particles of universe. Om! Peace! Peace!! peace!!! be there.

Such a great mind-set! In prayer for the welfare of whole universe, all abodes, forest, water, earth, space!! This Mantra is even beyond the consideration of the world population is one family. All this is belittling of even the "VASUDHEV KUTUMBAKAM".

It is surprizing why such norms of "susanskrit sanskar" are distanced by modern societies and cultures and the civilizations.

Author feels the Sanskrit norms be adopted by newer generations of the universe for the welfare of the world.

Taittariya Upanishad (Krishna Yajurved), Prayer Shanti Mantra: From – Brahmanand valli Anuvak 1.

Om Saha navvatu I

Sahanoau bhunaktu I

sahaveeryam Karva vahe I

tejsvi Navdheetmastu ma vidvishavhe I

Om shant! Shanti!! Shanti!!!

Teacher and pupil both, be protected, both eat together, both must shine together by studying, both must not be jealous of each other. Oh God peace be there.

AUTHOR'S SELF EXPRESSION

After reading GITA and its translations or listening discourses, multiple times and co-relating it with my research outcome and exposing the meaning of stanzas perceived, it was a realization that the knowledge expressed in question answer form with citation of examples, GITA is nothing but pure Science of Nature, soul and super soul – the Creator, the Creation and the creature. This book is my personal interpretation of stanzas and the hidden scientific meaning in the literature. Some may agree or some may disagree with the interpretation. Those who may not agree to interpretation, it is urged and requested to ponder and scrutinize the subject in detail, interpret and then come to conclusion. It is not religion, cult, faith or communalism (sampradaya). It is eternal knowledge (SANATAN DNYAN). Knowledge of self (ATMADNYAN) is discovery of vedik studies which is since time immemorial, even before known history. Surprisingly HINDU DHARM is not the word from Ved or Puran. This nomenclature is from invaders of BHARAT (India). After scrutinizing vedik literature I feel that I have not come across the word "HINDU" in Vedik literature. Enrolment of people in some group with assurance of liberation cannot be considered truly knowledgeable – ATMADNYANI (आत्मज्ञानी). VEDIK knowledge is essential for understanding the "self".

Soul is "self-sufficient" and needs no tout or commission agent to understand or perceive it. The mediators, can only be professional with self-interest. Ancient sages after understanding soul, never propagated any cult, faith religion or communalism (sampradaya). Action and work ability of soul was thoroughly investigated scientifically. Then only peoples were advocated by many ancient Sages, Rishis and Munis, since time immemorial (vide History of Gita chapter FOUR of this book). Hence it is considered as eternal. Modern religions are revolving around a particular person BUDHA, JAIN, SIKH, HINDU deity, CHRIST or Mohommad Paigambar etc.

"Sanatan Dharm" is Science of Nature, Soul and Super soul, creator, creation and creature. Author wishes, this fact is realized universally in its original sense. With this sense an attempt is made to explore science and true knowledge in GITA stanzas. The attempt is to reveal true knowledge in its truthful perspective and not with any intension to hurt any faith or religious group. If anyone feels hurt, author extends unconditional apology in advance.

This book is a conscise description about hidden science hence careful scrutinization of sentences and stanzas of Vedik literature will provide joy of understanding and enlightenment. Exploration of science has been presented in a limited manner. It is assumed that the reader will try and understand the secrets of the subject, after reading the book from beginning till end. Suggestions, if any for improvement are earnestly requested.

DR. B. G. Matapurkar.

MBBS, MS (SURGERY), MNYAS (USA), FAIS, FIMAAS

SCIENCE IN GITA

Indians have a boasting habit that everything discovered in modern time and science has already been known to their Sanskrit speaking ancestors. It can further be added that not only the scientific knowledge exist but scientists of past civilization were much more advanced than of modern time. Modern science is slowly approaching to the same goal. The truth gets unfolded when new modern discovery is explored. Viz. embryonic Stem Cell discovery and its human use is scientifically scripted in AADI PARV of MAHABHARAT. Growing babies outside uterus and outside human body. This is still to be discovered by modern science. Recently discovered science of "Particle Physics" was known to sages thousands of years back. Only patiently one must discover what is scripted in Vedik literature, understand and analyse it and co-relate with modern science with un bias mind. Author humbly summon all the youth of the world to ponder, scrutinize and attempt research on the fantastic knowledge in Sanskrit literature, for the ultimate benefit of the humanity.

Gita is not a Religious book. Understanding of spirituality is also a great scientific job to be performed. Spirituality or spirit or ATMA can be understood after the knowledge of material, physical NATURE. Hence every time a great stress is enforced in gaining knowledge on spirituality, in Gita, VED and PURAN, which are full of such knowledge and needs to be followed. It is a Psychoanalytical and highly scientific book. 18 chapters of Gita is basically conversation between warrior Arjun and illumined Lord Shri Krishna. It is a summary of Ved and puran to emphasize and explain to all aiming at Arjun. All this is to answer questions emphatically and assertively because the vedik knowledge was prevalent in general population. The knowledge is cited as examples to emphasize the answers. Hence, cannot be understood in isolation. Basic books like Shrimad Bhagwatam and relevant Sanskrit literature, must be referred for scientific explanations. In fact, Gita is a summary of Ved, Puran and Sanskrit literature. Therefore, the literature must be referred to explain hidden meaning and science in Gita.

Scientific facts are: Following is for general information:-

▲ Birth of Kaurav – 101 children developed outside the uterus and human body – a detailed version and scientific explanation has been given in a Chapter of Aadi Parv of Shrimadbhagwat Gita. 100 males and 1 female i.e. XX and XY pregnancy having a cleavage in between male and female in embryo. Aborted (after two years of conception, modern science knows it as **"arrested pregnancy"**). Aborted ball of flesh cooled, cleaned, treated (disinfected) and divided. Accidentally divided into 101 parts due to cleavage between XX and XY product of conception. Preserved in a safe and secure place (no details about secret place could be traced) for another two years. This is to reactivate dormant embryo cells for regeneration of fetal growth. The split into 101 embryo cells and kept in separate "kunds" containing clarified butter (word used is Ghritpurna Kundashatam Clarified Butter is author's interpretation). Details about herbs or chemicals are not mentioned. Kaurav's age is not separately mentioned (but Pandav's age is different) because the CONTAINERS (कुण्ड-Kund) were opened on the same day, one by one i.e. first opened is senior. In other word, the Sanskrit speaking civilization knew the art of developing fetus outside the human body of female in an artificial device/uterus.

- Life exists in plants – Shri Madbhagvat gita. Various life forms are divided as per their emergence and source in this universe – From eggs (birds, Snakes), From Seeds – (Plants and Vegitation), from Larvae –(Insects and Mosquitoes), and in Body form covered with membrane – (Animals and Humans)

- Embryology: Growth of fetus in uterus and development of body in uterus has been described in Shaktipeethank of Shrimadbhagwat Gita. Aiteriya upnishad (6000 BCE).

- Fertilization and chromosome in cells responsible for proper development has been described as GUNVIDHI – a proper name to carry forward the structure and function of cells. Sanskrit name is explanatory to the function of chromosome i.e. Cromosome = Gunvidhi – Shrimadbhagvat gita. (GUN= characteristics, VIDHI= procedure, arrangement).

- Energy and electromagnetic induction force responsible for maintaining structure and function of substances relating to Lanz's laws of energy and Faraday's laws of electromagnetic induction of forces. VAISHESHIK SUTRA BY MAHARISHI KANAAD.

- Development of Particle Physics by Maharishi Kanaad (7000 B C).

- Life of universe, rotation of galaxies – Bramha around Vishnu and Vishnu around Shiva has been mentioned in stanzas. Cosmos and interrelations of stars and planets, their dependence on each other. Shiva blessed Vishnu to protect Brahma and so on.

- Life span of Universe – Number of years have been figured out as life of Brahma, Vishnu and Shiva. 8000 Yugchatushka (= 4 yug together is 1 yugchatushka) = 1 Day of Bramha and similar Night). Like this 360 days = of Brahma's one year of BRAHMA, like this = 100years is life of brahma. Like wise life of Vishnu and Shiv.

- Embryo growth in vitro—outside human body – Shrimadbhagwat Gita. When the parts of fetus develop and in which month is explained in details. Heart beat in 2 mth of pregnancy and so on in – Shrimadbhagwat gita & Atreya upnishad.

🛦 Test tube babies from Sperm alone, ovum alone, from ovum & sperm both – Shrimadbhagwat Gita

🛦 Gravitational force – prashnopnishad, & Shankaracharya

🛦 Atom divisible – Madbhagvat, & Pudgalshatra (Jain), by Jain TIRTHANKARS.

🛦 Aryabhatt– knew value of Pi 3.14 centuries ago.

🛦 Space science. "Sphot vad". Big bang theory. Explosion is responsible for creation of manifest world. OM naad.

🛦 Other sciences were also developed by Rishis and sages of civilization of Sanskrit literature viz. some are mentioned in Mahabharat: Astrology, and Astronomy. Physiology. Zoology. Genetics. Science of Diet. Spiritual science. There is no mention of such things in Gita hence not discussed.

GENERAL CONSIDERATIONS

Author has not come across any book describing hidden scientific aspect exclusively in stanzas of SHRIMAD BHAGWAT Gita.

Saints of all ages and land throughout human history have been trying hard and struggling to reach the "Creator of Universe" i.e. GOD. The quest for truth and true illumination has resulted in different forms and names. Supreme reality ultimately is one and the same. This book tries to understand and explore the hidden science in stanzas of GITA, as per the existing modern science. No attempt is made to discuss the sect, faith or religion prevailing in modern human history. But the word mentioned as DHARM in GITA and available Sanskrit literature has been analysed and explained in the perspective of modern science. **Human history has witnessed the torture and insults of modern scientists like GALELIO and BRUNO, in the recent past, by the religious custodians having blind faith.** Currently with change of time, the same custodians now are eager to verify the principles of religion, scientific facts, in the light of modern investigations. But Vedik literature seems to be highly analytical and scientific in their approach in study of nature, Creator, Creation and creature.

Before exploring science in GITA, let us understand what science is and what is GITA? *Science can be defined as the promotion of natural knowledge, pursuing truth with the help of systematic investigation of the nature. It is based on the method, for the search of knowledge. The GITA in fact a narration in question answer form with suitable examples of prevalent knowledge and truth available and understandable to the contemporary society. Such illustrations are only to substantiate the narration and drive in the science concept from available Sanskrit literature.* Unfortunately, the Gita is misconceived as religious only. The real truth and true sense in Gita has been ignored.

The question answer form of Gita and scientific knowledge of common population has been narrated repeatedly in the text of this book. It is only to emphasize the needed fact. Kindly forgive the author for such repeated expression. Most of the time readers have the habit of reading the book not from the beginning till end. But read pages in between at randomly.

The reason for misconception about Gita is because of the word DEVTA used in Sanskrit literature and in contemporary human society it means GOD. The misunderstanding prevailed about GITA as religious book. In Sanskrit literature the elements, Nature, supernatural elements and substances, earth, Stars, Sun Moon etc. are respected as DEVTA. The word is also extended to Air (Vayu), Fire (Agni), Space (Aakash), Water (Jal), and many more supernatural substances. Read the word DEVTA in its correct perspective, then only one realizes that the Sanskrit literature is a study of nature, celestial elements, Soul, super soul, cosmic drama and so on. It is a scientific study of nature and supreme existence. (Prakriti and Prakriti purush).

"Science is a series of approximations to the truth. Scientific community of the world cannot claim to have reached finality. All theories and science facts are liable to revision in the light of new facts. This is both joy and inspiration of science that there appears to be no end to new knowledge and its interest. Each advance yields a more far reaching and interesting picture of physical world, while at the same time opening up fresh views in the shape of new problems awaiting solution." (A W Barton)[1]. Similar expression is in Sanskrit it is not the end literature exist. Sages used the word as Na Iti or Neti meaning, thus so far, we know further you explore. What a real scientific expression!

Exact translation of Sanskrit in English may not be possible, but attempt is made to provide near perfect meaning of Sanskrit word to English word, using Google search and Wikipedia. Even then, unintentionally, the meaning may not be perfect for which author requests for unconditionally apology, to all readers.

History of India is improperly projected due to vested interest of invaders in the past few centuries. The result of such history has created a doubt about RAMAYAN and MAHABHARAT whether these are facts or fiction. An interesting fact has come in limelight, in recent past regarding history of MAHABHARAT. From the pages of MADBHAGWAT a private search was arranged at a site where KRISHNA PURI – DWARKA was submerged in sea about 30 Km from of KACHHA (कच्छ), ex coast of GUJRAT (by archaeologist).

Marine archaeology has proved that the existence of the Dwarka and its submergence in the 2nd millennium B.C referred to in the *Mahabharat, Harivamsa, Matsya* and *Vayu Purans* (Sanskrit texts) is a fact and not fiction. Another crucial evidence is provided by *Mudra* (seal) in use of which there is a reference in the *Harivamsa.* Dwarka excavations by the archaeological department have demonstrated that the rich underwater cultural heritage of India can be successfully explored, excavated, retrieved and preserved.

(Dwarka to Kurushetra. Dr. S. R. Rao. Journal of Marine Archaeology (1995-96). Underwater Cultural Heritage. A.S Gaur and K. H Vora. Current Science Volume 86 No 9 May 2004. Further Excavations of the Submerged City of Dwarka. S. R. Rao. Recent Advances in Marine Archaeology Dwarka Mythical City Found Underwater. Weyland Yutani. www.unacknowledged.info)

In 2002 a TEAM OF NATIONAL INSTITUTE OF OCEON TECHNOLOGY PROJECT, in a routine photographic survey of the region, had photographed that area. Under the Leadership of Mr. KATHIROLI, Department of OCEONOGRAPHY, GULF OF CAMBAY was explored. MONTHS LATER DETECTED THAT THEY HAVE ACCIDENTALLY PHOTOGRAPHED A VAST ANCIENT CITY OF 7500 BC, SUBMERGED in 40 m deep sea. Dates proved by carbon dating of Fossilized log of wood from deep sea, this was published in India today magazine in 2002 February issue, THE LOST CIVILIZATION (**INDIA TODAY FEB 11, 2002**)[2]. **See photographs on page 4**

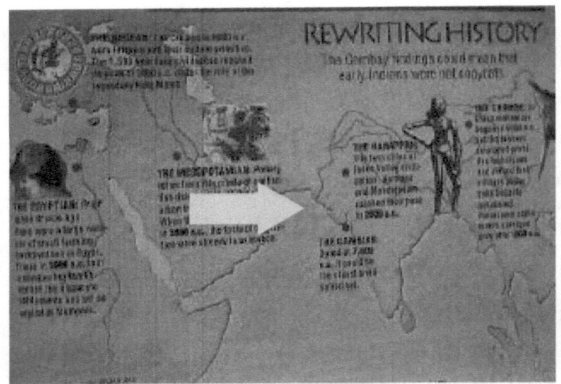 

Site explored 30 Km from Katchchha ex Gujrat coast. (Blue arrow)

Cover page of India Today.

Deep sea exploration. (India Today)

Another fact about MAHABNARAT, is that, on 13th day of Mahabharat war, after death of ABHIMANYU (son of Arjun), ARJUN vows to kill JAYADRATH (a warrior from Opposite army group of KAURAVS) that ARJUN would either kill JAYADRATH before sun set or immolate himself. Due to Solar eclipse sun disappears while war was going on, and fighting forces thought that the sun has set. Now as per the vow ARJUN will immolate himself. JAYADRATH and enemy forces started rejoicing, taking the advantage of the situation, ARJUN kills JAYADRATH. The total solar eclipses during that time were calculated by modern researchers. There is mention of two "Solar eclipses" during the war. One of them falls on 13th day of war. The knowledge about eclipses, was known to SHRI KRISHNA and one of the brothers of PANDAVAS named SAHADEVA, who had knowledge of stars and eclipses. After calculating, Dr. S. Balakrishna, of NASA, USA, revealed that the celestial phenomenon was a fact on that particular day of war[3,4,5,6].

From the above facts which sounds true, it can be considered that MAHABHARAT is less likely to be a fiction. It is a true and real history. Because of vested interest of invaders who ruled Bharat even changed name of BHARAT as India so that future generations forget Identity (Truly BHARATVARSHA the name of the country has been forgotten after few generations). There is no instance in the world when slave country's name has been changed. Textbooks in schools glorified invaders own interest only and ignored interest of inhabitants and the history of the country. It is well known that history is written by winner. Interestingly satellite images released from NASA indicates artificial construction under sea between RAMESHWARAM, a city of south India, and Shri Lanka, remnants of bridge built by RAM during RAM RAVAN war at the time of RAMAYAN period. (Remnants of bridge, site, photographs and, details about the history visit web site: http://en.m.wikipedia.org>wiki)

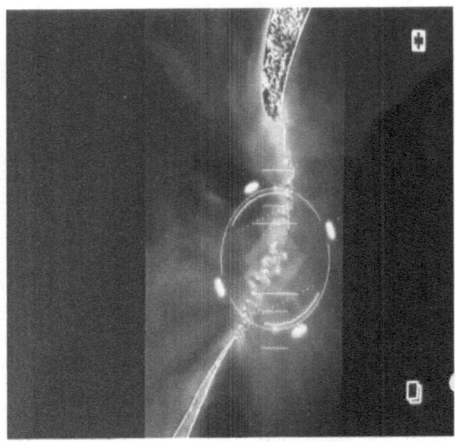 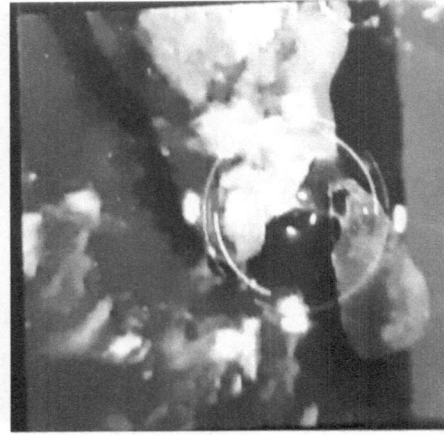

Manmade structure spotted by NASA.   **Satellite image of 'RAM SETU'. India and Shri Lanka. (Circle to highlight regions)**

Valmiki Ramayan. Yudha Kandam. 6.22.50-72. The bridge has linear alignment. It was a planned human effort. The bridge was built in 5 days – 14 Yojan on 1st day, 20 yojan 2nd day, 21 yojan 3rd day, 22 yojan 4th day and 23 yojan on 5th day. Total length 100 yojan and 10 yojan wide. (Ramayan, 6. 22. 68-72). Originally the name was "NALA setu" as built by an engineer NALA (नल). Later it was popular as "RAMA setu". Later on called as "SETU BANDHA". During Muslim invasion of Shri Lanka and ruled by AADAM. On his visit to bridge it was named as Aadams bridge. Under British it became popular as "Adams bridge". (Details on web site: www.bharatgyan.com).

The first ever universities of the world existed in BHARATVARSHA (now India). The university had multi storied building for rare books as library. The nine storied building housing 9-10 million manuscripts (Wikipedia), then how such a civilization be considered as illiterate? The library was destroyed and burnt by invaders.

In this connection author remembers a story heard about his great GRAND father. He had a collection of handwritten Sanskrit books preserved as his possession. The number could be about approximately 70–100. A foreigner, probably a German scholar collected those books on payment of Rs. 10,000/= at that time. The great grandfather thought that these are dust collecting articles and useless for him. The sum was a great deal for him. It must be a great treasure of knowledge.

A question can be asked can such a civilization be illiterate and ignorant or unscientific. It is hard known fact that, whatever begins on earth must end someday. The Sanskrit speaking civilization, which was once so glorious, having uncountable wealth, is sparsely existing at present. Sanskrit language having highest quality of grammar, scientifically developed phonetics, and considered as the mother of all the languages of the world. How then the civilization be considered as illiterate, orthodox and uncivilized? Presence of wealth has attracted invaders. Present generation has become a HINGLISH speaking (Hindi + English, a mixed language) civilization. That is the reality today. English medium schools are more popular in India today. Modern parents feel pride in admitting children in English medium schools. Sanskrit culture and social norms have been replaced by western culture and social norms of west. In India at present, the modern population is highly impressed by this so called modern culture and social behaviour. Even though it is a reality but at the same time it is a needed change when all wealth, literature, scientific facts have been looted or destroyed. Newer generation now to begin from scratch. This is a success story of invaders see Lord Macauley's statement in British parliament, in 19th Century. (Newspaper cutting in chapter three).

Purpose of this book is to explore the hidden scientific facts in GITA. After reading GITA stanzas and the examples cited indicate that scientific knowledge was popularly existing with general population during MAHABHARAT PERIOD. That is why examples are cited to explain the point of view of narration. Gita is a question answer between depressed Arjun and highly illumined Shri KRISHN or between a disciple and revered teacher. To understand those examples cited in GITA one needs to refer to main texts available elsewhere. This is the method adopted by the author in this book. Several books were consulted by the author like MADBHAGWAT, SHIVPURAN,

PANTAJAL YOG SHASTRA, other UPANISHAD WORK, VEDIK Literature, and modern Wikipedia and Google search etc. The relevant Verses are also quoted from such available Sanskrit books, to emphasize the point of view and explain the examples quoted, to authenticate SCIENTIFIC FACTS IN GITA.

Exploitation of nature by man is since time immemorial. VEDIK time RISHIS and sages, are not exception to this. VEDIK knowledge is ancient on earth. Not a single branch of science remained untouched by seers and RISHIS of VEDIK period. Highest of all sciences is spirituality. That is what GITA has preached in all its 18 chapters. How to keep environment clean by performing YADNYA, a sacrificial procedure in which specific material ingredients are offered to fire GOD, (basic material elements were considered GOD which are neither ISHWAR, CHRIST nor ALLAH etc.), in a unique chemical combination of specific materials in a particular ratio. The material ingredients in different combinations, cannot be without the knowledge of chemistry. It cannot be without experimentation. A stock of vast literature is available but in a hidden state. This book is an attempt to unravel scientific facts hidden in the stanzas. This book has envisaged the same aim. The literature of Vedic period is in Sanskrit only. **Unfortunately, majority of modern scientists do not know Sanskrit and majority of Sanskrit knowing pundit hardly know science, is a common observation.** This results in difficulty in exploration of hidden science in the Sanskrit literature. Modern trend is to understand Gita in its religious perspective. Science in Gita has been ignored. The modern youth is influenced by western achievements. Parents feel proud and glorified, if their children are educated in English medium school, of course which is the need of the day. Sanskrit subject is hardly taught in schools. At the same time students and their parente have hardly any interest for sanskrit subject.

Use of technique of RHINOPLASTY (Reconstructive nose repair) devised by ancient surgeon from Vedik period named SUSHRUT mentioned in SUSHRUT SAMHITA, is in use by modern Plastic surgeons. This indicates the advancement in cosmetic surgery by

Sage Sushrut. Not only this, it indicates their knowledge of blood supply and nerve supply of skin used for grafting. The CHANAKYA's ARTHSHASTRA is highly appreciated by modern scientific world. The knowledge of space, stars, and their constellations, science of Astrology, Astronomy etc. is a testimony that ancient India's intellectual capacity was also high. The great scientific knowledge of Sanskrit speaking civilization of ancient Bharatvarsha (India) be brought to the knowledge of world. At the same time indulge into research activities to explore scientific facts which are still to be researched.

After successful endeavour in his research on regenerating tissues and organs (US Patent), the author unravelled the mystery of birth of 101 children from Mahabharat book. There is scientific explanation in Sanskrit literature hat 101 children were developed in a device – as artificial uterus. This turned to be the triggering point to explore more scientific facts in SHRI MADBHAGWAT GITA. This book is an outcome of such endeavor.

Dr. Asha Matapurkar,

MB, BS, D C H. (Pediatrics), P.G.H.A.

References:

1.   Neo-organogenesis an Neo-histogenesis by Desired Metaplasia of autogenous tissue Stem Cells In vivo. A Critical and Scientific evolution with 125 years of review literature. American Society for Artificial Internal Organs Journal 2002. January issue of 2003.

2.   A new Physiological Phenomenon of mammalian body for Organ and Tissue Neo regeneration in vivo – Adult stem cell technology in the perspective of literature. Ind. J of Exptl. Biol. Vol.40, 2002 pp. 1331 to 1343.

FROM AUTHOR'S PEN

GITA has always intrigued me personally since time immemorial of my life. Reading book on GITA in Sanskrit, has also inspired my mind and soul even though I am not Sanskrit language expert. Every time I have read the stanzas from Shrimad Bhagwat GITA, I have perceived a different meaning. I have always felt that one reading of GITA is not sufficient to understand the meaning of every stanza. One must read it again and again. It inspires you differently every time. It is a sure shot path for success in everyday activities and actions. You need to meditate on the stanzas to understand the hidden meaning. Scholars of GITA from western societies have always felt that the Sanskrit speaking civilization is one of the oldest and great RELIGIOUS civilization of the world. Scholars have understood the importance of GITA preaching specially for modern society. Gita is considered as a religious book all over the world. It is interesting to know that the world RELIGION is a recent introduction in English during 18th and 19th century. Sanskrit word DHARMA is translated as religion. With exposure to words like Buddhism, Hinduism, Taoism, Confucianism, Christianism, Islamism etc. Meaning of religion, common consensus is belief in GOD, prayer and worship. The word DEVTA in Sanskrit is taken for GOD. In Sanskrit everything in universe like, for example: Sun as SURYA DEVTA (sun god), Fire AGNI DEVTA (fire god) VAYU DEVTA (Wind god), JAL DEVTA (Water god), AKASH DEVTA, PUSHAN DEVTA (Nitrogen god) and so on. In fact, all supernatural elements are named as DEVTA in VED, PURAN etc. by Sanskrit speaking civilization. Religion is understood as DHARM. Meaning of DHARM is totally different, it is basic quality of all living or non-living, material world or manifest universe. Correct word is "manifest" and not living or non-living, as everything is living in this universe. Everything in manifest material world has birth, growth, a life span and death including celestial being – stars, Galaxies and the whole BRAHM, VISHNU OR SHIV, as per understanding of Sanskrit speaking civilization. This fact is getting unfolded gradually by developments in modern science. Couple of centuries back, plants were not considered as living but now they are included as living beings. While the Sanskrit civilization has considered plants as living beings and named as VANASPATI DEVTA (plant god). The misunderstanding about the civilization needs to be revolutionized radically. How such a civilization be considered Orthodox? Invaders into BHARAT VARSH due to their vested interest tried to condemn this highly educated and scientific society as BUTPARASTH (idol worshipers), Religious (DHARMIK) and orthodox (ritualistic), cast society (real meaning is explained later in the kook). Contrarily the fact is otherwise. Basically my task in this book is multi-fold as I have to convince the world that DHARM is not religion in true sense in Sanskrit (http:en.m.wikipedia.org>wiki). The civilization was not an unscientific, idol worshiping nor ritualistic society. God in Sanskrit is considered as NIRGUN NIRAKAR (without any quality, Virtue less or formless) at the same time God is beyond time. In Sanskrit it is kaaltrayateetah.

Why I got interested in Gita? I have been reading Gita since my childhood. Philosophy of 2nd chapter has fascinated me all the time. I, as medical professional got interested in research of regenerating tissues and organs in the body after observing regeneration of tail in lizards. It was my belief that Laws of nature are intricate and camouflaged by revealed world of nature (MAYA) but have uniform applicability in nature. It was known that Plants once cut have capacity to regrow. Similarly, in animal kingdom lower forms of life have this capacity of regeneration of traumatized or damaged body parts. For example: Crabs, Lobsters can regenerate their damaged or severed claws. Lizards can regrow their lost or damaged tail. Similarly, some animals of lower life forms (as per zoology) also have the capacity to regenerate their tissues and organs. For example, Salamanders too can replace their tissues and body parts. If laws of nature are uniformly applicable, then why other living species like mammals and man cannot regenerate their body tissues and body parts. Multi-cellular animals Mammals and men have limited capacity to regenerate nails hair etc. Uniform applicability of laws of nature is explained below:

The illustrations below depict, the fact of uniform applicability of laws in nature. Photograph adopted from Readers digest Atlas. Based on the space knowledge.

(THE EARTH. By Arthur Beiser and the editors of TIME LIFE BOOKS. Hong kong, Life Nature Library. 2nd Edition, 1985).

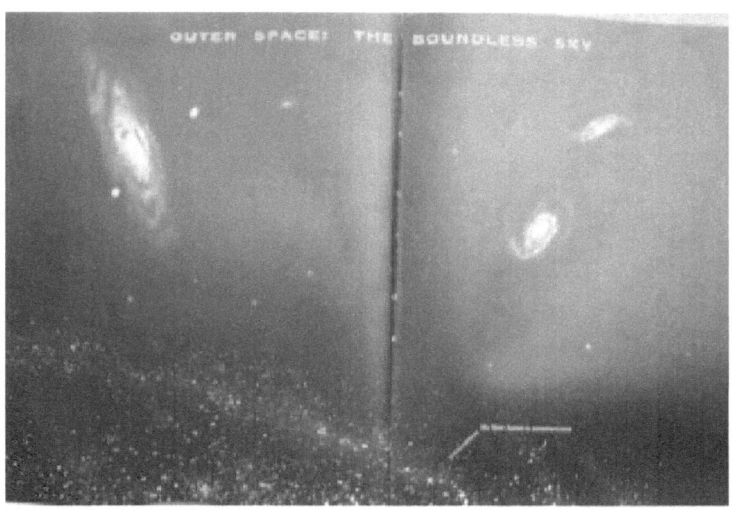

There are numerous galaxies in space.
Milkyway galaxy where Solar system exists. (arrow marked)

In space there are different Galaxies. One of the galaxies is Milky-way galaxy. It rotates around the centre of the universe. Red arrow shows site of Solar system in the periphery of Milky-way galaxy (fig above). The solar system with all the planets rotates around centre of milky-way galaxy. We all know that there are many galaxies in the universe, and these rotate around the centre of the universe. Our earth is part of solar system in our Milky-way galaxy. The solar system with all planets in Milky-way galaxy, revolve around the centre of milky-way galaxy. In figure below the Solar system with all planets including earth, revolve around central sun.

Solar system. Sun and Planets.

On the earth, the smallest known particle the ATOM, has Electron, Proton and neutron.

Electron and proton rotate around central neutron.

In conclusion, rotation of Galaxy, Sun, Planets, different factions of Atom are existing in the Universe i. e. the law of nature is uniformly applicable. Other facts about the uniform applicability of laws of nature are illustrated below:

Earth has water content on earth, which is ¾th water and ¼th land. Similarly, in comparison bodies of living beings on earth, the material body has similar composition that is 3/4th water and rest organic matter. The micro unit of animal body is cell. The cells have 3/4th water in their composition i.e. same as earth, or one can say composition water of earth, body or cell has uniform percentage.

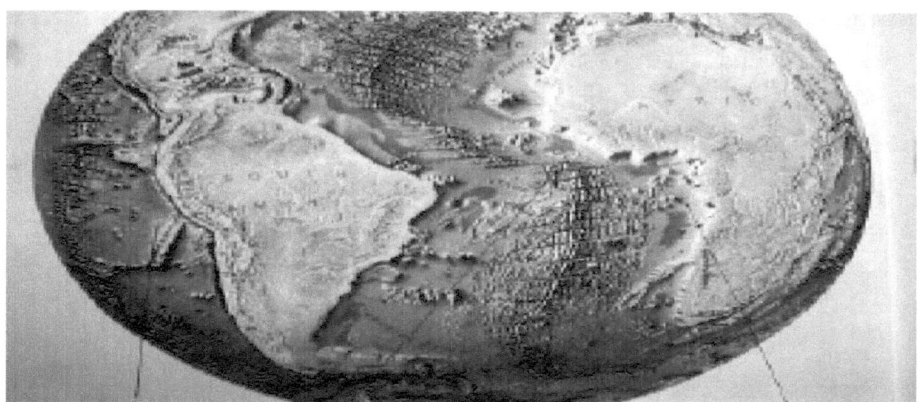

Earth - land and water relationship.

Earth has 3/4th water and 1/4th land so our body, and unit of our body the CELL also has 3/4th water. Such is the uniform applicability of the laws of nature. Apart from this if one compares universe and the multi-celled body, it can be realized:

In universe there are innumerable stars which take birth and die but universe perpetuates. In the body there are innumerable cells which take birth and die but body perpetuates. Not only this but whatever is in Universe it is in our body. Ved recognize presence of PANCH MAHA BHUT (Earth, Air, Water, Aakash and fire) and coined the word AHAM BRAHMASMIN or yatha brahmande tatha pinde. i.e. I AM THE UNIVERSE. This is a scientific truth as per modern science.

If laws of nature have uniform applicability, then why we cannot regrow our tissues and organs? To answer this question, one must understand the body formation in the womb. Our whole body is formed from the single celled fertilized ovum. The body has different tissues and organs, having different structure and different function. Mammals and men have lost their capacity to regenerate tissues and organs. This is because of specialization of tissues and organs for special function. For maintenance of smooth function of tissues and organs throughout life span the specialization of cells is essential. **The cells are unable to return to the fertilized ovum stage.** Therefore, the regeneration of tissues and organs capacity in specialized multi cell animals and men is lost.

Is there a possibility of manipulation of tissues and cells to help body of man and mammal to regenerate tissues and organs of the body? A simple and practical experiment on tobacco plant explains this fact. In plants especially in Tobacco plant manipulation is possible.

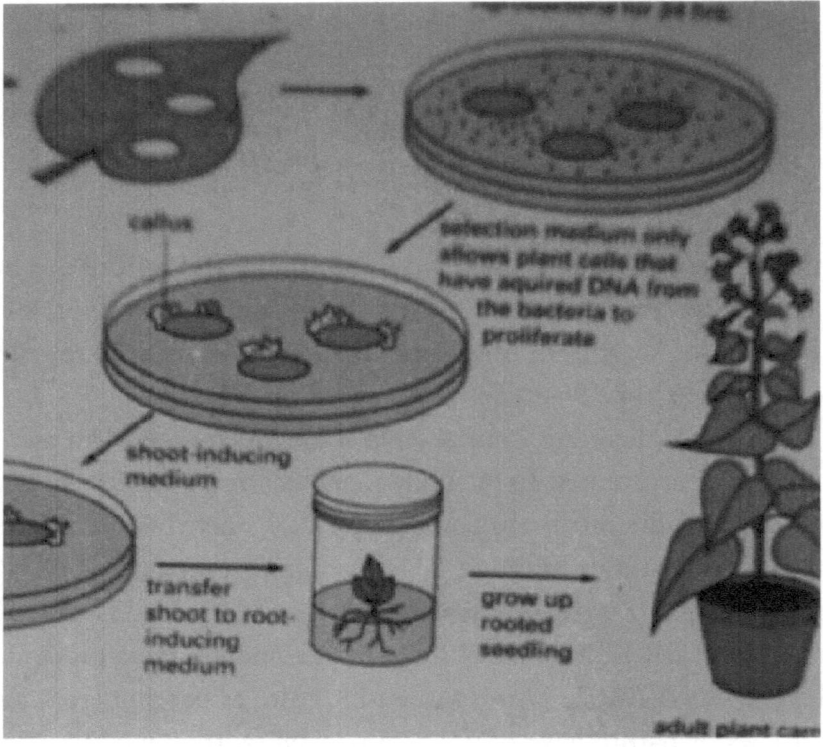

Manipulation in tobacco plant.

Plant growth and development can be manipulated. Cells of Tobacco leaf put into media containing agro-bacteria, in this media, the plant leaf cells use DNA of bacteria. Then in root and shoot inducing media roots and shoots start growing. A seedling is formed which grows into Tobacco plant. If laws of nature are uniformly applicable, then it is possible to manipulate regeneration of tissues and organs in mammals and man also.

As per the above illustrated description there is a possibility that tissues and organs in the body can be grown by manipulation in humans and mammals. Experimentation was attempted and could regrow successfully, a few tissues and organs in the body like aponeurosis of abdominal wall, Ureter, Fallopian tube, Uterus etc. in dogs and in near human models the monkeys (Vide International Patent from USA, Wikipedia.)

After getting success in such a venture the idea came to mind that how 100 KAURAV were born to a single woman. How a woman can give birth to 100 children during her life with a limited reproductive span. A female, on an average, has reproductive span of life or period from 15 to 45 years of her age. A span of 30 years of life, when conception is possible and children can be born. Normally 30 pregnancies are possible. A woman can give birth to 30 children. But during 9 months of pregnancy and 6 months of lactation, on an average, no ovulation takes place, so no possibility of pregnancy. Even with hypothetical imagination, if one considers that after every 15 months one after another without any gap the pregnancy takes place with quadruplets every time, only 94 children can be born during 30 years of reproductive span of life. Therefore, the MADBHAGVAT Puran was scrutinized thoroughly to find any possibility of some explanation written in the book of MADBHAGWAT PURAN.

Astonishingly and amazingly a full chapter is available in AADIPARV section of MAHABHARAT epic which has given reasonable scientific details that, how 101 children were born to GANDHARI from aborted embryo of GANDHARI, (wife of DHRITRASHTRA).

1. *The MAHABHARATA, Vol 1, Adi, Sabha, Aranyaka, and Virat Parvans, Bhandarkar Oriental Research Institute, Poona 1971. 107. 3 –page 156-159*
2. *Mahabharat Aadi Parv, Ch 107, page 583-587. By Dr. S.D. Satawalekar. (Mahabharat Aadiparv, 107, page 583-587,*

Quoting Sanskrit stanzas from AADIPARV section of MAHABHARAT: JANMEJAY –(king) asked question to (RISHI) VAISHAMPAYAN to narrate how and after how much time 100 sons born to GANDHARI (wife of DHRITRASHTRA ruling king of HASTINAPUR).

> ***Katham putra shatam jadnye Gandharya dvijsattama I***
> ***kiyata cheiv kalen teshamayushchya kim param.***

RISHI VAISHAMPAYAN explained to JANMEJAY that after conception GANDHARI continued pregnancy for two years without delivery. She became worried and depressed.

> ***Samvatsaram dvayam tam tu Gandhari garbh Samhitam I***
> ***apraja dharyamas tatsta dukhmavishat I***

When she heard about that KUNTI has given birth to children, she was forced to think about her pregnancy. Grasped with grief, without the knowledge of DRITARASHTRA, with great effort she forcefully aborted the products of conception. The product of conception which was for two years in her uterus fell down as fleshy ball, hard like iron.

> ***Tato jdnye manspeshi lohashthilev sanhata I***
> ***dvivarshasambhrutam kuksho tamutstrashtum prachakrame I***

In modern science, in medical subjects of Obstetrics and Gynaecology, if a pregnancy remains for two years without further growth, is termed as **arrested pregnancy**. The toxic products of arrested pregnancy may harm mother. To protect mother from the effect of toxic products of pregnancy the products of conceptions are entombed in a calcified covering. It becomes like a hard ball (लोहाष्ठीलेव). It is termed as **OSTEOPODIUM.**

On knowing this RISHI DVEPAYAN quickly reached there and saw the fleshy ball. He asked GANDHARI, why she has done this? She explained her desire to RISHI like this:

Jyeshtham kuntisutam jatam shrutva ravisamaprabham I
dukhen paramenedamudaram patitam maya

After listening, the birth of a sons to KUNTI, (wife of PANDU – co brother of GANDHARI) with grief she attempted abortion. You had blessed me with a boon of 100 sons. In place of 100 sons this fleshy ball has born. RISHI assured her that the same would happen. He ordered for 100 pots with purified butter and cleaned the aborted fleshy ball with cold water and cooled the product of conception – fleshy but hard like ball of iron.

Ghritpurna kunda shatam kshipramev vidhiyatam I
shitabhiradibharashthilamimam ch parishinchat

While sterilizing and cleaning the ball it was divided into small pieces. The number became 101 in due course of time.

Sa sichyamana ashthila abhavachchhatadha tada I
angushthparv matranam garbhanam prithgev tu

After this the newly created pieces of products of conception, transferred to containers filled with clarified butter ordered previously.

Tatastansteshushu kundeshu garbhanavdhe tatha I
Swanugupteshudesheshu raksham cha Vyadadhattatah: II 21 II
(Mahabharat., Aadiparva 107, 21).

Meaning: swa = self, Anushashit = controlled, Gupteshu = secret, Desheshu = place, Raksham = well protected, Vyadadhattatah = without harm

The containers were shifted to a secret secured place The containers should only be opened after two years. On decided time the containers were opened one by one. First opened was DURYODHAN – eldest

KAURAV. Along with 100 sons one born was a female child named DUSHALA.

Scientific explanation:

Scientific aspect of all this narration can be analysed scientifically as follows:-

The aborted ball of flesh was fertilized ovum with arrested further growth called as arrested stage of pregnancy. It must have grown to a certain level of embryo growth and then could not grow further for some reason not described in Sanskrit text. Arrested growth of embryo can result in arrested pregnancy. Embryo grown to whatever stage contains embryonic stem cells having "toti-potency" (total potential to grow whole body) in such cells. Such cells can grow into full and perfect, complete animal or human (e.g. CLONING of DOLLY). Embryonic stem cells can survive for a long period, like plant seeds. Example of Chickpea or Gram seed can be quoted here to explain the whole phenomenon. If Chickpea (chana) seeds are kept in a container on a shelf of kitchen, it remains as seeds for a long period. In the seeds the growth is dormant, if proper environment is absent i.e. moisture and soil etc. The germination is in suspended form in such seeds. The Altered environment of these seeds like providing soil, moisture etc. it starts growing shoots and roots **after a lapse of some time.** This period is needed for cells to acquire level of germination. The embryo was split and while splitting it got divided into 101 pieces as described in Sanskrit text. In an embryo if it contains male and female foetuses – both (XX and XY pregnancy) then a cleavage appears in between the two. That is why probably accidentally the pieces became 101. The female child born from manipulation was named as DUSHALA. She was a sister of KAURAVS.

Secret and secure place or containers are not explained in Sanskrit texts. How the nourishment to growing foetus was provided? The literature is silent on this. I could not place my hand on any of details about this. The GHRIT filled containers mentioned in literature could not be analysed in detail as no record for that could be traced.

With such a vast scientific detail, in Sanskrit in MADBHAGVAT PURAN, the scrutinization and meditation on the stanzas of Gita a scientific BOX of PANDORA was opened for the author which is presented in this small book. There may be many more science detail author has missed other scholars can add with their understanding. It will be a tribute and honour to SCIENTISTS (Rishis) of Sanskrit speaking civilization.

Science is **study of and knowledge about the physical world and natural laws.** The same has been described in GITA as examples to satisfy questioner. In scientific knowledge – proof and seeing is believing that confirms and grants approval for success in scientific achievements in manifest material world. VEDANT, apart from this accepts believing is seeing as it brings to light certain things which otherwise go un-noticed. Greatest example to this is GOD. Spiritual science was developed. No one has seen the soul or super soul. It is a belief, which forced Rishis, seers and monks of Sanskrit speaking civilization to develop transcendental meditation. They could achieve un-parallel success by providing knowledge like YOG, travel with micro-body SUKSHMA SHARIR to distant planets etc. (see modern day planetary travel with micro-body, by Dr. Vartak 's experiments in advance to NASA's Mars, Jupiter exploration. Page 8-15.
(Upanishadanche Vidnyannishth Nirupan, by Dr. P V Vartak, Part 1, publisher Vartak Prakashan, Vartakshram, 497 Shanivarpeth, PUNE 411030.)

Science believes in material world dimension of life. What is perceivable through the sense of body of material world. There is no other dimension to life in scientific materialism and therefore it is changing from one point to other viz. previously it was believed that the earth has flat shape and sun rotates around earth. This belief has changed with progress of scientific developments.

The dreams experienced during sleep, are real but on awakening turn out to be unreal. VEDANT philosophy concludes that the material world seems real like, as in dream but may not be real. Similar thinking that in deep sleep one is without conscious thought in mind and creates pleasantness, happy feeling on awakening. **VEDANT philosophizes that the thoughtlessness of mind in meditation and SAMADHI, certainly provides permanent happiness (CHIDANAND). It sounds a great philosophical significance.** On similar philosophical note Lord KRISHNA has mentioned in GITA Chapter 8, stanzas 16, 17, 18, 19, that the material world and the whole universe manifest and un-manifest repeatedly in a cyclical manner. (attention of reader is drawn to "time scale" of chapter 8.). It further mentions that because of this YOGIS believe in un-realness of material world and PURUSH TATVA or Super soul as eternal. YOGIS believe in TIME FACTOR (KAAL TATVA – कालतत्व,) as dimension. Super soul is beyond time. The word CHITTA has no equivalent meaningful word in English language, hence "MIND-SET" is used in the text.

A brief description of Mind-set (CHITTA) is as follows:
The mind-set has been described in nine different states in YOG SHASTRA. In yoga philosophy the CHITTA/Mind set has an important place.
(Patanjal Yog Pradeep 4, Kaivalyapad, page 134-135, Gita press Gorakhpur, ISBN 81-293-0011-7)

These nine states of "mind-set" (CHITTA) can be achieved by practice. This helps highest ever achievements in material world. Philosophers and scientists have knowingly or un-knowingly have reached such mind-set before their results and scientific achievements. Yog philosophy in YOGSHASTRA has expressed in words AND IN A SCIENTIFIC MANNER, in 4 different sections: SAMADHI PAD (51 TEXTS), SADHAN PAD (55 TEXTS), VIBHOOTI PAD (55 TEXTS), AND KAIVALYA PAD (34 TEXTS). These are the various stages of progress to ultimate liberation from bondage of material world.

I feel Sanskrit speaking civilization had achieved perfection in all branches of sciences known today. In some spheres even the RISHI and sages were more advanced in scientific achievements as compared to achievements of modern science. Not only this but such scientific facts were known to common people of the civilization. Therefore, examples are cited in question answers of GITA between Student and teacher.

To understand the original aspect of mind-set (CHITTA), nine groups of "state of mind" are described:

1. **Awake-state (JAGRIT AVASTHA).** Senses are in wakeful state and work normally in live material body. Proof/evidence (PRAMAAN), rearrangement/reversal (VIPARYAY) option/rational choice (VIKALP), Memory/fluctuations in mind set (SMRITI VRITTI), all are in awakened state aroused and working.

2. **Dream state (SVAPNAVASTHA).** Mind set with the help of mind, work in memory state. With the help of senses, micro body works in dream state (material body remaining still). The person appears and projected working in mind-set and appears working in dream state.

3. **Deep sleep state (SUSHUPTI AVASTHA).** Mind-set is totally unaware of any physical knowledge. All work stop in micro-body. Material body in deep sleep state. Soul is in similar state.

4. **Procedural state (PRALAYA AVASTHA).** Mind-set is in a state of deep sleep. The difference is that the mind set in singleness is in deep sleep while mind set in totality is in holocaust (प्रलय). All living entities are in a state like deep sleep.

5. **Initial SAMADHI.** The mind-set is in concentration state. Frequency of concentration is high. Soul is in similar state. Quivering mind set starts concentrating.

6. **Higher conscious state** (SAMPRAGYAT SAMADHI, EKAGRATA). Reality starts revealing. Material body and senses stop working. But micro-body actively starts concentrating in conscious state. Soul enters into similar state.

7. **Conscious state – Right knowledge state (Stage in between** SAMPRAGYAT SAMADHI and **A**SAMPRAGYAT SAMADHI **– vivek khyati).** Mind differentiates between soul and mind-set in micro-body.

8. **Individuality loss state – Pious mind set (ASAMPRAGYAT SAMADHI-SWARUPAVSTHITI).** Individuality or "I"-ness is lost. Soul is in super soul state.

9. **Unification of causative factors of mind set and "mind set – state". (PRATI PRASAV AVSTHA).** Mind-set unifies with its cause. The soul enters into super soul state.

CENTER FOR SCIENTIFIC WORKSHOP
(THE BATTLEFIELD KURUKSHETRA)
LABORATORY FOR STUDY OF ELEMENTS
(DHARMKSHETRA)

Abstract: The words used in the chapter in Gita are misinterpreted and mis-understood. The real meaning as per the Sanskrit literature are analysed with quotations from the literature are provided to authenticate the real meaning of words like Dharmkshetra and Kurukshetra. An attempt by author, is made to put forth the real meaning. After ignoring the prevailing religious tinge to Gita and analysing on scientific ground, the meaning is presented in this chapter in its original perspective.

Before the discussion on Chapter one a brief knowledge of history of war of Mahabharat is mandatory for better understanding of Gita – a scientific song on Creator, Creation of nature and Creatures of manifest material universe. A greatest philosophical and scientific dialogue just before war between Learned disciple and revered and illumined teacher. The story is of a conflict between KAURAVS and PANDAVS. Both happened to be cousins. Both belonging to the KURU dynasty. Both descendent of King BHARAT – a ruler of earth. The word Maha Bharat has been derived from King Bharat. The intricacy of the story of conflict starts due to the fact, that DHRITARASHTRA and PANDU are brothers from KURU dynasty. Elder of the two brothers was Dhritarashtra, born Blind and hence throne came to younger brother – Pandu. But Pandu died early. The children of Pandu were reared up under the care of Dhritarashtra, and he became the ruler for the time being. Both cousins received military training from expert teacher DRON acharya, with counselling from grandfather BHISHMA an expert in warfare knowledge. Hate and envy prevailed between the cousins as students. DURYODHAN – eldest Kaurav – with consent of Dhritarashtra, plotted to kill Pandavs but protected by Shri Krishna, who is nephew of KUNTI, the mother of Pandav. Later Exile for Pandavs was planned for 13 years and one year of anonymity (Adnyatvas) as YUDHISHTHIR lost gamble which was rigged by Kaurav group, and Pandavs were cheated in the game of dice. On return from exile and anonymity, Pandavs demanded the Kingdom which was refused. The war became inevitable.

The opening line in GITA is very interesting. Gita is first ever scientific philosophy developed which is fundamental to whole mankind, on the battlefield – just before inevitable "world war" between rival cousins, (world war as warriors from all the available countries of that period participated as rival forces). This field has historic and Vedik importance and a place where scientific fundamental Dharmik (not religious) and pious rituals performed for the benefit of mankind and hence nick named as DHARMKSHETRE.

Dharmkshetre kurukshetre samveta yuyustavah I
mamkah pandavashcev kimkurvat sanjaya II 1 II
(Gita Ch 1, Stanza 1).

What scientific details are hidden in above 1st stanza from Gita. DHARM is not as modern religion it will be explained later. Name and place are both significant. It signifies the mind of DHRITARASHTRA about the ultimate victory of his sons (KAURAV). As per the laws of nature and the constitution of UNIVERSE, nature favours the pious and rightful deal. KURUKSHETRA as per VEDAS is a place of worship (scientific workshops). Dhritrashtra was doubting the influence of the **holy place** on the outcome of the war. He knew that the war is between VIRTUOUS and EVILS.

Dhritarashtra and Pandu belonged to "Kuru dynesty". KURU was the ancestor of Kaurav and Pandav. King Kuru belonged to BHARAT dynasty. Kurukshetra name is derived from the name of King KURU. At the same time KURU in Sanskrit means to start action or do action. Kshetra means field. **The field of action is KURUKSHETRA.** Geographically, Kurukshetra is region bet**ween two rivers – Saraswati and Drishadwati.** It is of utmost importance to note that **King Kuru supported "Research and development" in this region for prosperity of the region, land and people. He performed YADNYA, translated as Sacrifice and misunderstood as religious practices as offering material things to fire god.** ARCHAEOLOGICALLY it is proved that Samrat Ashok the great, made it a centre for learning. In ved, Kurukhetra is also a place of sacrifices for DEMIGODS as well – DENIZENS of heaven (Bhagwat Gita as it is, by Swami Prabhupad, A C bhaktivedanta, Founder of International Society for Krishna Consciousness). Scientific proof of this statement (DENIZEN) which could not be ascertained – probably history of King Kuru can be traced to King of "Jambu dveep PRIYAVRAT" whose son Aagnidhra ruled Jambudveep. His eldest son NABHI later handed over the reign to RISHABHA. His son Bharat later ruled southern VARSHA (Varsha means – region – desh – देश) part of Himalaya. That is how this region is called as "Bharat varsh", where Kurukshetra is a city. The ancestors of Bharat known as Demigods who performed actions to gain scientific knowledge and spirituality for the society.
(The details are in Shri Ling Maha Puranank. No. 1. Year 86. Gita Press Gorakhpur. Chapter 47, Stanzas 1-23.)

In Sanskrit literature yadnya is not offerings to fire god. YADNYA is action with a purpose. Let us understand the true meaning of yadnya. By performing YADNYA many people achieved SIDDHI. MIMANSA and VEDANT DARSHAN (मीमांसा और वेदान्त दर्शन)

clearly stated in KARM KAAND about work and actions. The work and actions can be desired action (ISHTA KARM – इष्ट कर्म) – is YADNYA. The PURTA KARMA is social work. In PURVA MIMANSA first statement is "अथातो धर्मजिज्ञासा" i.e. now discuss DHARM. All work and actions are included in KARM-YADNYA. Rishi MANU says i.e. by YADNYA can manifest body (ब्रम्ह शरीर). Is it possible by offering material articles to fire GOD? To elaborate this further, let us understand ASHVAMEDH YADNYA, a type of YADNYA which means as per modern understanding is "**army march past**" in the region with decorated HORSE (ASHVA), followed by army. The regions or countries where the horse goes unopposed that region is included in the king dome of YADNYA performer. If any region opposes the horse movement, war and fight is mandatory to claim the region. Such YADNYA was performed by Shri Ram, king of Aayodhya. His sons named Lov and Kush took away the horse. There was a war between Shri Ram and Lov – Kush (Story from Ramayan – A story of king RAM).

Sanskrit literature has described 5 types of main YADNYA. These are as per the Manusmriti:

Adhyapanam brahma yadnyah pitruyadnyastu tarpanam I
homo devo balirbhoto nrityayadnyotithipujanam II 70 II
(Manusmriti. 3. 70).

1. BRAHM YADNYA. ADHYAN: Study (perusal of topics and subjects), ADHYAPAN: Teaching (Tuition) etc.,
2. PITRUYADNYA. Food and water offerings to departed souls of forefathers. How the departed souls get the food and water is different topic beyond the purview of this book. (Explanation is available in Sanskrit literature – MANUSMRITI 3, 76).
3. DEVYADNYA: Regular worship and sacrifice for GODS.
4. BHUTYADNYA: Food for the animals (pet and community).
5. MANUSHYAYADNYA: Respect of guests and visitors (coming to you).

These yadnya are for the householders (GRIHASTH). Because as a householder (GRIHASTH), one performs acts and deeds due to which numbers of insects (life), or jeev are killed or harmed during daily chores of activities. As a penance these YADNYA are performed.

This means YADNYA is not just sacrifice of material or food, offered in sacred fire. **YADNYA is action with a purpose.** Confusion is due to use of word DHARM in the perspective of modern religion (see MAHARISHI KANAAD in later chapters). Interestingly many other types of YADNYA have been described. All are beyond jurisdiction of this book. In "Taittariya upnishad" (in Brahmanandvalli, Anuvak 5. Stanza 1) a stanza very clearly declares the meaning of Yadnya:

Vidnyanam yadnyam tanute I
karmani tanutepi cha I
vidnyanam devah sarve I

Brahma jyeshtha mupasate I
vidnyanam brahmachedved I
tasmachenna pramadyeti I
sharire papmno hitva sarvakamansamshnute iti II 1 II
(Taittariya Upanishad 2. 5. 1).

Science is study of Nature and Natural things along with knowledge about them. Knowledge spreads the feast of sacrifice and feast of work. YADNYA means which gives Knowledge or one gets science and knowledge by work, sacrifice or Yadnya. Nature and PRAKRITI is from BRAHM. It (BRAHM) is senior – worth the worship. Or Vidnyan is BRAHM. YADNYA facilitates Knowledge. One who knows this will be sinless. This means YADNYA means research and study of scientific facts for ultimate human benefits.

Author wants to emphasize the true meaning of YADNYA. In short YADNYA is for acquiring expertise or "SIDDHI". To attain Siddhi action and study of the essential subjects. This is sacrificial attempt to gain success or siddhi, which cannot be without performance of action and work to yield results, but when performed with dedication to god its outcome is without bondage. This was the mental attitude of Oppenheimer father of Atomic bomb (discussed later in details). Same is reiterated in GITA in chapter 3 stanza 9, says:

yadnyarthat karmanonyatra lokoyam karmabandhanah I
tadartham karm konteya muktasangah samachara II 9 II

(Gita 3, 9).

Gita further says (Chapter 3, stanza 10, 11, 12) the work and actions be performed together (highlighted part of stanza below). Perform YADNYA to produce food – the most essential for life and live life happily. Actions to please the GODS –i.e. elements (all supernatural like RAIN, AIR, AGNI, AAKASH etc. are GODS).

Sahyadnya prajah srishtava purovach Prajapati I
anen prasavishyadhvmesh vostitvshtakamudhak II 10 II

(Gita chapter 3, 10).

Ishtanbhoganhi vo deva dasantye yadnyabhavitah I
teirdattanpradayeibhyo yonbhunkte sten av sah II 12 II

(Gita 3, 12).

After completion of action and work successfully godly elements will provide all necessities of life to Yadnya performers. Scientifically performed attempts certainly leads to success. This meaning is acceptable if religious tinge to yadnya is kept aside. After creating living creatures creator – Brahma said (Purovach prahapatih), the provider of facilities certainly will help you.

Therefore, the correct meaning of YADNYA be taken as work and action for achievements and knowledge for betterment of self and society. In Sanskrit the meaning of YADNYA is actions for acquiring knowledge – attempt to gather knowledge. DEVTA and elements perform sacrifices YADNYA i.e. actions and reactions to develop new elements in nature as stated in later chapters. Prajapati – Brahma performs Yadnya to create nature.

The battlefield named Kurukshetra, is also labelled as "DHARMA kshetra" because of innumerable YADNYA performed to study the "Sanatan Dharm" on nature, soul and super soul, creator, creation and creature. Because of this Kurukshetra has historical and Dharmik (not modern religious) importance. Because of performance of numerous scientific work shops and studies by ruler King KURU, for knowledge, the place is named as KURUKSHETRA. He performed multiple Yadnya. The cultivation of spirituality and overall knowledge of nature – Prakriti and Purush – self or Aatm for the ultimate good of the society, the place is also called as DHARMKSHETRA (not modern religious). Later on the SAMRAT Ashok – the great declared this region as centre for learning. It is a sacred place of pilgrimage from Vedik time. This battlefield where the greatest ever philosophy was delivered by teacher to disciple before war – a unique combination which is never happened before nor likely to happen in future. The name "Dharm Kshetra" is just and appropriate nomenclature of this region.

What is DHARM as per Ved and Sanskrit literature? Maharishis KANAAD's version is:

Dharyati iti dharma

Elemental assumption of virtues by each and everything in the manifest or unmanifest nature is DHARM of that being. May it be element, living or inert existential in the universe. In Aiteriya upnishad it is cleared that "Aatma dharma" or "virtuous soul" has its own DHARM:

koymatmeti vayamupasmahe I
ktarah s aatma I
yen va pashyati yen va shrunoti yen va gandhajighrati yen va
vacham vyakroti yen va swadu ch vijanati II 1 II
(Aitareya Upnishad 3. 5. 1. Aitareya Upnishad, chapter 3, Khand 5,
stanza 1).

Virtues of soul are with which one sees (pashyati), one listens (shrunoti), one smells (gandhajighrati), one expresses word (vacham vyakroti), one understands good or evil. These are the dharm of soul. VED is Knowledge (or dnyan), this Dharm is VEDIK DHARM. Gaining knowledge is dharm of Ved. Name HINDU is given by invaders of BHARAT (India) to bring it at par with other contemporary religions which were developed in past 2000 years or so. One can understand that an eternal, Sanatan dharm cannot be compared with changeable contemporary religions. This was the reason that VEDIK SCIENTISTS did not use any nomenclature as HINDU, BUDHA, Muslim, Christian, and JAIN etc. Because of scientific developments, actions to gain knowledge (yadnya karma), study of Nature, Creature, Creation and Creator at KURUKSHETRA the region is considered as "DHARMKSHETRA". This study was continued even after Mahabharat by ASHOK the great, King Harsh etc. There is evidence that a Chinese Traveler HIUEN TSIANG – a Budhist Monk, has visited this place because of its scientific, historical, Dharmik and Vedik importance.

Another scientific question is how SANJAY could see the battlefield while sitting in the room of DHRITARASHTRA?

Sanjay was a student of RISHI VYAS. With his blessings and boon, he had a great vision (DIVYADRISHTI) by which he could see exactly the happenings on the battlefield. Exactly as we observe on TV, a modern scientific devise. MADBHAGWAT GITA is silent on details about this issue. But in Bhagwat many descriptions are there about BOONs and Blessings. KUNTI, mother of Pandav was blessed to have children from celestial beings like SURYA, INDRA etc. No scientific description could be traced on this subject. The man who could see future – NASTRADAMUS is known in present history as DIVYADRISHTI or a devise like modern day television existed? One can only speculate if a nuclear war bursts and all is destroyed then after thousands of years of evolution from stone age may not see any of the modern age devises – computers, TV, Laptops, Pen drives for computers, rockets, space craft etc. Man visited space and walked on Moon surface, will only be an imagination and no proof will be available. All will perish. What will remain is analysis on history by archaeologists.

DHRITARASHTRA's mind disclosed here that his son DURYODHAN and others will be wiped out and pious, rightful religious PANDAVAs headed by YUDHISHTHIRA would be established by LORD KRISHNA. This is the significance of the words DHARMAKSHETRE and KURUKSHETRE apart from the historical and VEDIK importance. Unfortunately, modern translations and discourses all revolve around modern religion (Not SANATAN DHARM).

A wonderful psycho-analytical philosophy of GITA follows in the following verses. Arjun requested Shri KRISHNA to take the Chariot in the centre of two rival enemy forces, ready for the war. So that he can see who all have gathered in the battlefield. The evil-minded sons of king Dhritarashtra, and others, who sided the evil mind of DHRITARASHTRA and associated as enemy forces. Interestingly Arjun has been referred as GUDAKESHA (GUDAKA is sleep and who conquers sleep is GUDAKESHA, who conquers sleep and ignorance). After seeing his Kith and kin, brothers, grandfathers, teachers, friends, father in Laws etc. he became overwhelmed with compassion. His limbs started trembling and mouth drying up, skin burning up. His body trembling body hair erecting (hair standing out). Bow slipping from his hand. All signs of nervousness. He was astonished to see their fighting spirit. This is not due to his body weakness or physical/mental weakness or subdued fighting and killing spirit. This was due to his soft heartedness and devotion to humanity as humble devotee of God and good qualities of mind, as well as material qualifications. These symptoms and body phenomena occur during great spiritual nervousness, ecstasy or great fear of loss of life and fear under material conditions. Arjun's Deep affection for community and family members, expressed here due to his natural compassion for them. The other symptoms of slipping bow, burning heart and skin are indicative of material conception of life. "Why should I kill relatives who are ignorant and greedy, or they kill me in exchange of even three worlds" (TREILOKYA). Would it not be a sin to kill relatives and friends? Arjun feels that, what pleasure he would derive from killing the sons of DRITRASHTRA, his own cousins, friends, teachers, in-laws, relatives, etc. How then the killing of all these persons, the post war life would be enjoyable or – the kingdom and post war happiness, would be enjoyable. His envisioning of painful reverses of the battlefield was responsible for the condition. He expressed to Lord Krishna for refuge in Jungle and live a life of frustration rather than killing his kinsmen. Arjun was a saint warrior and politician and saint human being. For a common man the six types of aggressors need to be killed as per VEDIK injunctions. Attacks with deadly weapons, setting house on fire, one who gives poison, snatchers of riches, occupying other's land, one who kidnaps one's wife, are considered aggressors. Arjun was not a common man. State administrator should be saint but not coward. An aggressor in RAMAYAN era, RAM was saintly but not coward. RAVAN was aggressor but RAM gave reasonable and enough time and lesson which is unparallel in the history of mankind of the world. Therefore, even today the population wants to live in RAMRAJYA (kingdom ruled by Shri Ram). In MAHABHARAT era the case was different because enemy to ARJUN was his own relatives, friend, teachers and in laws – persons as aggressors. Hence, he thought to forgive them. In fact, Arjun was moved compassionately seeing his kith and kin, and discussed with Shri Krishna with same emotional mood. Unfortunately, in present time in India, the same situation is prevailing. The Indian nationalism is at crossroad with secularism and pseudo secularism. In Gita stanza 28 and 29 emphasizes the same fact:

drishtvemam swajanam Krishna yuyustum samupshtitam ll28 ll

Sidanti mam gatrani mukham ch parishushcha sharire me romharshashch jayate ll 29 ll.

This is Indian mind developed through centuries of ethical cultivation, not found elsewhere in countries of world. Gita in fact has described the Indian mind in Mahabharat period and present time also.

Kshatriya is not supposed to refuse for the battle or gamble if invited. But Arjun's mind was of different perspective. He considered that if other party may be blind even if enemy as a relative, which would be of evil consequences, so no need to accept the challenge. Obligation is essential if outcome is good but if otherwise one cannot be bound. So, he considered not to fight.

According to modern science the stand is PSYCHOANALYTICALLY correct. Another explanation post war effect could be destruction of dynasty and families etc. provided in Gita 1. 39 is more scientific:

Kulakshaya pranashyanti kuldharmah sanatanah I
dharm nashte kulam kritsnamdharmo bhibhavtyut II 39 II
(Gita 1, 39).

Destruction of dynasty, family traditions destroyed, and family members become irreligious. Non followers of laws of nature and SANATAN DHARM. The spiritual evolution of younger members gets destroyed if elders are absent. If irreligion is prevalent in family and society, women are polluted, and degraded, and misled into adultery resulting into unwanted progeny which prevails in society. This devastates the welfare activities of the society. The scientific explanation is as follows:

Om tadsaditi Shrimad bhagwatgitasu upnishadsu brahmavidyayam
yogshastre shrikrishnarjun samvarderjun vishadyogo nam
prathamodhyayah

Upnishads are in fact preach philosophy of Ved. Gita is upnishad of similar nature. Practical philosophy of life and tendencies are displayed. Analysing the Arjun's tendency here, in this stanza, scientific facts can be explored. Genetic tendency depends upon DNA. Quality particles in DNA decide the tendency. Tendency is the resultant of years of continuous cultivation of groups good/bad virtues in mind-set of persons. Such tendency provides AURA of the person. The aura can be perceived by YOGIS. Aura can be photographed by KIRILIAN photography camera developed by a Russian scientist. The bright light emitted by the physical body of a person can be photographed by the Krilian camera. The light tinge or hue changes with different tendencies. Sages, Rishis of Sanskrit speaking civilization after research decided the tendency from aura and named it as VARNA. After years of study, expressed the society on VARNASHRAM style, depending upon the tendency emanated and perceived from the auras or varn of persons of the society. Fallen females give birth to wild and uncultured children. The progeny in turn derange the social discipline. This is the scientific meaning of stanza 39 of Gita mentioned above. Modern science accepts that breed and tendency can change with the tendency administered by the parents at the time of conception. Arjun was aware of this scientific genetic fact hence expressed it in these stanzas. The SANATAN DHARM and VARNASHRAM DHARM all designed to enable human beings to attain the ultimate salvation and all-round progress of society and nation. Class system and duty related social system (VARNASHRAM – people groups in – BRAHMIN, KSHATRIYA, VAISHYA, SHUDRA) is inevitable in human society. This system is politically unnecessarily highlighted by the rulers everywhere, especially in BRITISH INDIA for divide and rule. At the same time slavery of generations, over superiority and ego by some higher casts led to chaos in the society. A simple question needs answer about this VARNASHRAM system. Can queen of Britain, from 10, "Downing street", of Great Britain sweep the roads in the society? President of USA will ever cook for public or society? Then why condemn the scientifically based VARNASHRAM system of Sanskrit speaking civilization. This was duty related, work oriented and AURA and tendency of persons – VARNASHRAM system of SANSKRIT SPEAKING CIVILIZATION. It needs no condemnation but on the contrary intelligently and properly applied with reasonable justification for the betterment of society. Mindset, intelligence, skill in work, education of persons, all decides for the category of work in the society. Elaborately described in later chapters. the woman conceives nurtures and develops fetus with emotions of human society, in Uterus for 9 months. The mother is considered as symbol of creation, culture, and language for the society and the nation. In the same context tendency of KAURAV can be compared as 101 children developed in an artificial uterus outside the mother's body. Foetuses developed outside mother's body are deprived of human emotions for two years.

Note about Kirlian Photography:

Foot note One of the opinion also be noted (Kirlian photography was accidentally discovered in 1939 by Semyon Kirlin and his wife Valentina after a visit to Krasnodar hospital. There, the pair witnessed a patient receiving high-frequency electrical generator treatment, which led them to conduct some experiments of their own.

During the trials, they noticed that objects placed on a photographic plate and subjected to intense electric fields create an image on the surface. Because the technique is a type of contact print, there is no such thing as a Kirlian photography camera. Still, you can use a transparent electrode instead of a plate and capture the images with any modern digital camera.

The Metaphysical and Supernatural

Due to the nature of the discovery, as well as the Kirlians' own beliefs, the process has come to be associated with the metaphysical and supernatural, as well as an important part of New Age philosophy. The term 'aura photography' is often synonymous with Kirlian photography, and there is a wide variety of mystical healing practices that are labeled as being 'Kirlian.'

Naturally, a lot of myths related to the subject emerged over time. And just like it is the case with many other spiritual and paranormal practices, there are plenty of believers who are ready to vouch for it, in spite of science having disproved their convictions. So, where do facts end and where does fiction begin in the case of Kirlian photography? Let's find out.

Kirlian photography might be fun to produce and look at, but it is by no means a supernatural event. It does not capture auras, and it doesn't have anything to do with the metaphysical. It is merely a natural reaction of photographic film to the corona effect of high voltage electrical energy, which makes it a imple light trick, – R. Gaspari. PIXI 2018)

Regarding the mind set of Arjun before the war, under such a great psycho-analytical mind set, described above, Arjun, decided to give up fight to salvage humanity from destruction and unnecessary genocide on both sides of assembled rival forces on the battle field of

KURUKSHETRA. In the same context the mindset of scientist who is father of Atom Bomb, changed.

(THE GITA OF J. ROBERT OPPENHEIMER written by JAMES A. HIJIYA).
Richard Rhodes. "I am become death": The agony of J. Oppenheimer". American heritage w28.6 (Oct. 1977).

OPPENHEIMER considered as father of ATOMIC bomb. He as director of the laboratory that developed atom bomb. It was tested in desert of New Mexico at a site named as TRINITY TEST SITE on 16th July 1945. Oppenheimer's reaction to the explosion was "I am become DEATH – YAMA the god of DEATH, destroyer of worlds. GITA scripture came to his mind. The agony of Oppenheimer's mind is hidden in these words. Later he quoted Lord Shri Krishna – AVTAR of VISHNU, addressing Arjun's mental depression before the MAHABHARAT WAR. Oppenheimer was a scientist to create scientific developments is his duty as a scientist. Use of such a deadly weapon when, where and why, is the duty of politicians and the decision is entirely a prerogative of decision – making persons. When Oppenheimer went to Harvard (1922-25), he began seeking a more profound approach in HINDU Classics. He acquired a deeper knowledge in BHAGVATGITA. He studied SANSKRIT. In fact, his mental condition full of compassion and depression like Arjun in Chapter one of Gita. His duty was only to invent as per his DHARM as scientist. Gita in fact solaced his mind set – CHITTA (चित्त) and ATMA, after the fire ball of atomic bomb test at TRINITY test site, Maxico. Trinity is in SANSKRIT is united three gods as TRIMURTY composed of three gods BRAMHA –the creator, VISHNU – protector and SHIVA – the destroyer. In fact scientist followed his DHARM as a scientist and rest left to politicians. Thus, his sinful effects of work done automatically negated. This has nothing to do with religion. The same is the preaching of GITA. How solace provider is GITA to Oppenheimer, one can understand.

In chapter 1 of Gita ARJUN discusses the ill effects of war with friend and relative Lord KRISHN, his friend and charioteer, guide. If one ponders on the questions and concern of ARJUN, a great saint warrior, one can conclude that Arjun as Saint, because such persons only can think of evil effects of killing of one's own family. The crime of destroying family and friends is avoidable. **The devastating effects in future** of killing own family and friends in war, which cannot be enjoyable. It is the greed of fighting forces that is preventing evil minds to recognize fault of fighting with family and friends. The ill effects of such war are described in the stanzas (GITA 1, 37–44): Arjun's knowledge on subject of genetics, clearly indicates how deep was the knowledge existed in the civilization of that time. Reader is refered to Sigmund Feud's psychoanlutic theory and principle. (journals.uchicago.edu)

Rival forces were aware of the relatives and friends as enemy, but due to greed unable to appreciate destruction of dynasty, family traditions, SANATAN family DHARM (not religion but family mind set) and future irreligious activity and behavior of forth coming progeny and generation.

(Kulkshaye pranashyanti kuldharma sanatanah Gita, 1, 39. See page 24).

Female folk of such family get polluted and adulteration prevails. Once female folk gets degraded, the unwanted progeny is inevitable. Such unwanted population make family as hell to live with. The ancestors of such family fall down as liberation (मोक्ष) of such population becomes difficult or non-existent. VARNASANKAR means cross bred children. Children born from undesirable mother and father. Children not wanted by parents. Children grow up to be nuisance in the world. Consequently, the whole world is in chaos. Crime rate goes up in society. Society is not peaceful and safe. The progeny becomes Cancer for the society called as VARNSANKAR.

Adharmabhibhavat Krishna prdushyanti kulstriyah I
strishu dushtasu vashneya jayate vrnasankarah II 40 II
(Gita I, 40).

All kinds of community traditions and projects, family welfare activities are destroyed and devastated. Family traditions destroyed the population always dwell in hell (नरके नियतं वासो). This is the result of sinful actions of war with desire, greed and attachment to royal happiness at the same time material gain in material world, from the success of war.

Sankaro narakayev kulghnanam kulasyas ch I
ptanti pitaro hyesham luptapindakodakakriyah II 41 II

(Gita 1, 41).

Unwanted distorted progeny (sankar) the traditions of society are not practiced due to which the PINDODAK KRIYA is not followed. Hence forefathers dwell in hell. Why Arjun need tell all this to illumined and knowledgeable Shri Krishn? What is PINDODAK KRIYA offering food to departed soul? What is science behind this philosophy of PATANTI PITARO – falling of forefathers in hell? This indicates the common knowledge of people of that period had knowledge of genetics as well as the transformation genes resulting into distorted progeny.

If the birth of KAURAV is considered, who were developed outside mother's womb in an artificial devise – "GHRITPURN KUND SHATAM" (described in "From Authors Pen" page 18-37), where the progeny was devoid of Mother's thought and vibrations inside the womb and the progeny became emotionless and greedy without any consideration towards society or humanity.

In 43rd stanza of chapter 1., it is clearly narrated that it is traditionally heard from teacher to pupil information or disciple succession, that when family traditions are destroyed the, people always dwell in hell.

Utsannkulgdharmanam mnushyanamjanardana I
nnarake niyatam vaso bhavatityanu shushrumII43II

(Gita 1, 43).

That is how knowledge was prevalent in the society. The public was aware of scientific development during that time. Under such mind set of Arjun, he further concludes that it would be better even if DURYUDHAN, son of DHRITRASHTRA, the ruling king, with arms in hand, kill Arjun while unarmed and unresisting, on the battlefield. This decision taken for salvaging the welfare of the society.

Ydi mampratikaramshastram shastra panyah I
dhartrashtrara rane hanyustame kshemtram bhavet II 45 II
(Gita 1, 45).

The stanzas that describe the genetic knowledge in the vast and available literature of VED and PURAN are discussed in later chapters.

Stanzas 39-45 stated above Gita exposes us to philosophy of GARUD PURAN. Which deals with departed soul its existence and ultimate liberation etc. This was the Vedik tradition which believed in SHRADHA and PINDODAK KRIYA-Gita 41, above. It is basically homage to departed soul. Garud Puran describes journey of soul after death. The science of SHRADHA philosophy contemplates the use of mental power and mind-set. As per Gita the deeds and work in current life decides the travel and next incarnation of soul of departed material body. But SHRADHA performance on earth by surviving relatives of soul help and provide solace to departed soul. It is comparable to helping and directing the space craft or space station from earth station using electromagnetic force signals. The science of SHRADHA can be believed if one accepts that the soul exists. Sanskrit speaking civilization with vedik knowledge believed in "absolute truth" philosophy and a fraction of which is soul in material body. At death of body the soul departs. The spiritual energy in the body enlivens the material body. This energy can be photographed by KIRILIAN photography. The AURA of body or light photographed remains even after death, confirmed by photography (Cut leaf or amputated leg or arm also shows complete aura of leg on photograph). This remains for some time and then deciphers. GARUD PURAN believes that the soul hover around body after death up to 13 days and then goes in space. Hence SHRADHA is performed on 13th day, this is PINDODAK KRIYA. This is mystic and clair (Clairvoyance – clear vision – French word) or intuitive knowledge. Yogis have this capability. This is getting accepted by modern science gradually.

CHAPTER TWO

SCIENCE OF SANKHYA YOG
SPRITUAL, SOCIAL AND PHYSICAL SCIENCE
(SUMMERY OF VED AND PURAN)

Abstract: Traditionally thinking person as Arjun, is convinced by the then Scientifically oriented philosophy of SANKHYA YOG by the author of Gita. The tendency of Arjun is in common man even today. The Sankhya yog was prevalent in the society at that time. Dogmatically thinking suppositions of common man thinking, in evil fearing humans is still existing. This is the mob psychology of common bewildered man. The same can be influenced and removed by Sankhya Yog philosophy. The philosophy is explained scientifically in the chapter.

This chapter is named as SANKHYA YOG (Science of Yog Shastra):

Om Shrimad Bhagwat gitasu upanishadasu brahmvidyayam
yog shastre shrikrishnarjun samvaderjun sankhya yogo nam
dvitiyodhyayah

Science of spirituality is "Sankhya yog". It is explained in later chapters of this book at appropriate place. Tendency of Arjun at this juncture of war was not to kill the relatives, friends, teachers, and innocent people of society or at the time of using weapons on such fellows would be impossible for Arjun to win war. With such a mental attitude of fighter as enemy forces are not alien. With such a fickle mind decision for good or bad for the society was impossible. Hence Lord Krishn took the help of SANKHYA philosophy. The state of mind of Arjun, compassionate, depressed, eyes full of tears, being a Kshatriya, such a conduct was not expected of him. Arjun wanted Krishna to kill the DEMON OF MISUNDERSTANDING in him who has overpowered him in discharge of his duty as KSHATRIYA. The mental and physical status of Arjun surprised KRISHNA. Krishna asks Arjun, "where from such impurities which were not expected from a person of ARYAN heritage belonging to civilized class of men with spiritual realization". Aryans believed that the aim of life is realization of ABSOLUTE TRUTH.

Na tve vaham jatu nasam n tvam neme janadhipah I
na chev na bhavishyamah srve vayamatah param II 12 II
(Gita 2, 12).

This means Shri Krishna says that "you (Arjun) and me (Krishn), and all assembled persons are existing now were existed before also and shall be existing in future. At the outset it is difficult to accept this concept. In other words, Shri Krishna says that the body is perishable and inside these bodies there is eternal element –The soul. It seems confusing outright. To understand hidden scientific fact in above stanza details need be referred to other source of Sanskrit literature. To understand this let us understand **KATHOPNISHAD Chapter 2, section 2, stanza 13.**

nityonityanam chetanaschetnanameko bahunam yo
viddhati Kaman I
Tamastmstham ye nupshyanti dhirastesham shantih shashvati
netresham II 13 II
(KATHOPNISHAD Chapter 2, section 2, stanza 13).

One knows that the material body is perishable, it is not eternal. The soul or energy which enlivens the body is a fraction of supreme soul. The Super soul which energizes all which is present in universe, is eternal. One who knows the inner self, the wise who see the super soul get SAT CHIT AANAND – true satisfaction and eternal peace. Body takes birth and dies. Energy-CHETANA that enlivens senses. This has been explained in chapter 1, on page 25. (Aitareyopnishad 3. 5. 1):

Koymatmeti vayamupasmhe I
katarah sa aatma I
yen va pshyati yen va shrunoti yen va gandhanajighrati yen va
pashyati va swad cha swadu ch vijanati II 1 II
(Aitareyopnishad 3. 5. 1).

The meaning is what is soul? By which one sees, listens, smells, or tastes is energy that enlivens body without which body is dead, everything in universe, sun, moon, fire, atom etc. exist due to that "soul-energy". In absence of which everything is non-existent. This is what Krishna has expressed in that stanza 2. 12, stated above.

Gita here indirectly propagate the theory that humans are immortal. Why Lord Krishn speaks that Ajun and KRISHN existed before? This means Shri Krishna says that "you (Arjun) and me (Krishn), and all assembled persons are existing now were existed before also and shall be existing in future. How this can be explained scientifically?

Let us analyse a stanza from Taittariya Upnishad:

Athah deivihi I
triptiriti vrishto I
balamiti vidyuti I
yasha iti pashushu I
jyotiriti nakshtreshu I
prajatirmritamananda ityupsthe I
srvamityakashe II 21 II
(Anuwak 10, Bhrigu valli, stanza 21 from Taittariya Upanishad).

Divine mandate or command of nature – satisfaction due to rain the Sanskrit word is VRISHTO, – VRISHTI it includes more elements than only rain, strength in electricity, success in animal activities, energy in star constellations, Amrit – like pleasure in praja – population in reproductive sense organs and all in Aakash. All these are almighty's commands. Bold and underlined part of stanza above indicates praja or population continuity provides AMRITANAND with the organs of reproduction. The AMRITATVA means no death element, i.e. species generation continues is mandate or command of nature like other elements mentioned in the stanza and translated above. Scientifically it can be explained by modern cell science – cytology and it's chromosomes or in Sanskrit GUNVIDHI. Propagation of species is effectively carried out by reproductive organs. It is mandatory command of nature and going on from generations to generations. It needs male and female in – Multicell organisms. It is a problem not limited to two but three as the third one is due to be produced. As one understands that the third one has a tremendous influence on the body of mother where it is developing. It does not depend upon any thing but transforms body where it has started dwelling (mother), to its advantage, and prepares mother's body for nourishment even after delivery by developing milk in the breasts. In short the whole process is NATURE dependent (DEIVY, see stanza above – Athah deivihi I). This way Amoeba – single cell becomes two and propagation continues. The plants propagate through seeds. Similar way it is in multicell organisms. The population continues. The microelement for this is DNA. It is in each, and every cell. With use of this DNA the nature propagate progeny. The species achieve immortality. The DNA is made up of organic matter (or earth, water, fire, wind and space). The population is basically a product of this which is immortal till the life span of Brahma and later the supreme soul as Sanskrit literature states. This is the hidden science in the stanza by Lord Krishn to Arjun. That is why Lord Krisna informs Arjun that 'The enemy forces and all in the battlefield were present before and will be present later as well'.

What modern science consider about soul? With the help of KIRILIAN photography soul energy which enlivens the material body, the scientifically soul energy has been supported to some extent. Vedik literature is full of facts about this energy. The soul possesses faculty of knowing, perceiving sensations of pain and pleasure. Modern experimental psychology discovered and knows the electrical counter part of soul known as "TAIJAS SHARIR". Human beings are born with electric charge (about 500 Volts). This does not change with age but changes with the type of character the subject is developing. This indicates AURA around living beings (plant, animal and human beings). After death, this energy leaves material body but dissipates in the surrounding. This indicates the energy or soul hovers in atmosphere surrounding body. This is CONFIRMATION OF TAIJAS SHARIR EXPRESSED IN JAIN PHILOSOPHY. Taijas sharir is electromagnetic energy – part of soul. A link between outer body and KARM SHARIR which is outcome of matter and spirit. In fact, "taijas sharir" is a coat of luminous matter over "Karm sharir" which forms aura (Kirlian photography).

In modern scientific language, the mind and matter activity make "super radio" waves. Body's all cells send different waves to numerous brain cells for tuning. Incoming of these waves is influx of subtle Karmic matter. One can call Karmic matter as 4th state of matter. Other three states of matter are Solid, Liquid and Gas. Activity of good character attract meritorious matter while bad character attracts opposite Karmic matter. Karmic matter in body is responsible for moving soul from one body to other, and soul remains in confines of universe, due to gravitational forces. One who sheds off Karmic matter, the soul follows the path of liberation. The soul being the lightest, moves to the top of the universe and remains there as divine. It is unable to move further as no ether a medium for movement. This is probably the scientific aspect of VEDIK liberation.

To elaborate further, Kathopnishad is useful. In Chapter 2, Valli 2, stanza 15, it says:

Na tatra suryo bhati na chandratarakam nemo vidyuto bhanti
kutohyamagnihi I
tamev bhantamnubhati srvam tasya bhasa
sarvamidam vibhati II15II
(Katahopnishad 2, 2, 15).

In chapter 15, Stanza 6 of Gita similar statement has been expressed. Abode of supreme soul is where sun, moon, fire, fail to illuminate. Everything exists and illumined by "self-manifest" supreme soul. Supreme soul is not dependent upon sun or moon light but on the contrary it is otherwise. Once reached there never returns. Similar stanza is in Gita 15, 6.

From above stanza it is made clear that everything in manifest universe is existing due to the super soul. The super soul is collection of tremendous energy. **Basically, it is TEJAHPUNJ and is a unique source of energy which provides energy to whole of the universe.**

To understand the **"absolute truth"** one must refer to main book of SRIMAD-BHAGVATAM (1.2.11 and BRAMHA SAMHITA – 5.1. Lord BRAMHA himself has concluded that KRISHNA is the supreme personality. No one is equal or above him. He is BHAGVAN (primeval Lord). He is the cause of the causes. This concept is difficult to appreciate. It needs explanation. ABSOLUTE TRUTH is explained in SHRIMADBHAGVAT. It is realized in three phases of understanding by the knower of absolute truth (highlighted words in stanza below):

Vadanti tattatvidasttvam yajdnyanmvadvyam I
Brahmeti Parmatmeti Bhagvaniti shabdayate
(SHRIMADBHAGWATAM 1.2.11).

Above stanza speaks of 1. Brahma, 2. Parmatma and 3. BHAGVAN. To explain:

1. Aatma (BRAMHA),

It needs little explanation here. "AATM" fact is very nicely described in ISHAVASYOPNISHAD SHANTI MANTRA:

Om purnamadah purnamidam purnat purna mudachyte I
purnasya purnamaday purnamevavshishyate I
om shanti! Shanti!! Shanti!!!.

BRAHM is "whole" or "complete" at its base level. From the whole only whole can be generated. A thing in higher state of material existence is not whole i.e. fail to generate the whole. But still has BRAHM or energy in a different state. It is in transformed state of specialization or material state in different quality. This can be explained by example of STEM CELL which is building block of living beings. Living beings which can be unicellular or multi-cellular:

Scientifically speaking microscopic living beings – Viruses and bacteria and even some unicellular creatures LIKE AMOEBA, have the power to divide and multiply with perfect replication of itself and produce progeny. These are whole and complete in themselves. These cells have senses to protect and find food for themselves. In multi cell organisms, which are grouped into male and female, they are not complete as they fail to produce similar organism like unicellular organisms. Male and female unite to produce progeny. But they do have BRAHM/energy in their constitution. These multi cell organisms also have cells in them in primitive level as primordial cells, recognized by modern science, in the form of "stem cells". These multicellular animals have different tissues and organs for different functions e.g. liver, kidney, brain, heart etc. Such tissues need special cells for the specific function. The cells get specialized for special function as need of the body of multicellular organism. The specialized cells are unable to return to stem-ness and fail to generate similar cells. Specialized cells form specialized tissues and specialized organs (as per modern science) for special function. The stem cells can produce exact replica of themselves and multiply. But can also maintain stem ness in them, i.e. when cell divide. one cell maintains stem character but other specializes. If whole is taken out from whole the whole remains. That is what the UPNISHAD expresses in the stanza. Many more examples can be quoted in this regard. Success of author's research (on regenerating tissues and organs in body) is due to this BRAHM fact described thousands of years back by Sanskrit Speaking Civilization. How can such narration be unscientific? Only reason is that we fail to interpret it properly. This book is an attempt in that direction. Similarly, the whole BRAHM – the universe, can be explained. Sanskrit speaking civilization had experienced the fact that, Shiv, Vishnu and Brahma are essentially responsible for generation, maintenance and destruction of the whole universe.

Namo rudray haraye brahmane paramatmane pradhanpurushay
sargasthitya karine

(SHRILINGMAHAPURANANK, Gita press Gorakhpur, year 86, number 1. page 25 – 26. Editorial),

Meaning of highlighted words = (rudray = SHIV TATVA, haraye = VISHNU, Brahmane = BRAHM)

= Brahm (sarga = generation, STHITYE = maintenance, ANTE = end),

sargasya pratisaragasya sthiteh karata maheshwarah |
sarage ch rachaya yuktah satvasthah pratipalane |
prati sarage tamodriktah sae v trividhah kramat |
adikarata cha bhutanam sanharata paripalikah |
(SHRILINGMAHAPURAN - 4, 35-37, page 26)

Creation, maintenance and destruction all is done by SHIV. Execution done with three forms Rudra-SHIV, VISHNU and BRAHMA **(Scientific Details are given in 10ᵗʰ chapter of this book).**

Shiv – is self-manifested, existing or generated by itself (SWAYAMBHU). As per SHIV PURAN, Shiv is a TEJAHPUNJ (Heap or collection of COSMIC CONSCIOUSNESS-CHAITANYA) as stated above. It is a Fire ball – a ling of AGNI which cannot be measured in length, breadth or thickness, it is extremely vast and endless. Repeated blasts are taking place in SUN as modern science knows. With every blast in sun, solar flares are generated. Similarly, blasts are taking place in the AGNI LING. Some are thrown in the form of fire balls of burning gasses. On cooling become galaxies and forms BRAHM. AGNI LING is surrounded by hot primordial gasses. VISHNU is in the form of PRIMORDIAL gasses and remain in dormant phase depicted as human figure sleeping on serpent bed, floating on KSHIRSAGAR as ocean with whitish hue in appearance. This is for understanding of common population. Scientifically speaking, SHIV TATVA basically is without form, shape or virtue. In fact, BRAHM SHIV element (tatva) is the main reason of whole material world. (For details visit 10th chapter of this book – Taittariyopnishad). Basic five elements – Earth, water, Light (TEJ, PRAKASH), Air, and Space (aakash) are derived from SHIV TATVA, responsible for creation of material world.

2. **PARMATMA is super soul and**

3. **BHAGVAN is supreme soul.**

The DIVINE aspect can be explained by example of "SUN" which has three different aspects. SUNSHINE, Sun's surface and the Star itself.

Student of **sunshine** is preliminary student. One who understands **sun's surface** is advanced student, while one who enters the planet itself is the highest form of student. Sunshine is the effulgence in nature and is only brahma feature of material world or fraction of **absolute truth**. Knowing sun disc is PARMATMA feature of absolute truth. A student who enters the heart of the sun and understands the personal features of the SUPREME ABSOLUTE TRUTH. It is the BHAGVAN feature of the truth. Subject matter is the same i.e. The sunshine, sun disc and inner affairs of sun itself. But the student of the three are at different planes and are not in the same category. Sanskrit version is like this:

Vadati tat tatva vidasa tatvam yaja dnyanam adavayam I
Brahmeti parmatmeti iti shabdayate I II II II
(SHRIMADBHAGVATAM I.2.II).

Ishwarahparmah krishnah sachidanandahvigrah I
anadir aadir govindah sarva karankaranam II I II
(BRAMHASAMHITA, 5.I).

The chapter infuses in the minds of reader that self – realization is by analytical study of material body and the spirit soul. The realization is possible when one works without attachment to the outcome or result and no attachment of any kind in material world. The kind of compassion and depression, agony in mind, would not lead to higher plane, but to infamy. The supreme person SHRI KRISHNA questions Arjun, where from this dirtiness and lamentation came upon ARJUN, at this time of crisis.

Kutastva kashmalamidam vishame samupsthitam I
Anaryajushtamswargyamkirtikaramrjun II 2 II

(Gita 2, 2).

ANARYA are the persons who do not know the value of life and such mind may not lead to higher planes of life. Arya means person with progressive thought in mind. A person who is keeping un-progressive thought in mind is ANARYA. It has nothing to do with race. Arya and Anarya are words used for good or bad tendencies. Invaders of Bharatvarsha (name of India before invasion) due to vested interest projected Arya as race and considered Arya as Invaders of India to justify their invasion. Lord Krishn condemns the behaviour of Arjun and discards his thought as unsocial, weakness of heart, lowest category attitude, and detrimental to the person, society, and whole nation.

VEDIK philosophy teaches us that the aim of life is to search for absolute truth. As explained above, it is good to practice the BHAGVAN level of the absolute truth. The GURU of Shri VIVEKANAND, Shri RAMKRISHNA, was of the opinion, that GOD is far, far away from the minds engrossed in worldliness, involved in satisfaction by sense enjoyments and worldly success of material world. Intellectuals having worldly intelligence creates pride and doubts with pride in wealth, learning etc. This leads to a turmoil in hearts responsible for more doubts. Childlike faith is essential in mind. A child believes instantly without any doubt what-so-ever Mother narrates to child. No doubt on that. Childlike faith is different from BLIND FAITH. Blind faith is due to fear and need of security, a desire for acceptance.

Eternity of soul confirms the presence of soul in different forms of bodies. Scientific aspect of stanza of Gita 2. 12 stated above is rival forces on the battlefield were present before and likely to be present in future as well.

Arjun lost his conscience as Warrior because under the effect of confusion, of purpose in life. Weakness of his behaviour due to fight with teachers, friends, cousins, relatives etc. Confusion in thought process due to DHARM – constitutional duty, as Kshatriya and his attachment towards his relatives in heart and mind. Surrendering to spiritual illumined friend Lord KRISHN, Arjun requesting for guidance.

Karpanyadoshopahatswabhavah pruchhami tvam
dharmasamudhachetah I
yachhareyah syannishchitam bruhi tanme shishyasteham shadhi
mam tvam prapannam II 7 II

(GITA 2. 7).

Miserly weaknesses of Arjun's mind have confused the ideology of life. Perplexity prevents decision making efforts of a person. Arjun requested for a decisive opinion to get rid of his mental confusion.

Before one understands the meaning of and science of chapter two, one must clearly unravel the meaning of "DHARM" word thoroughly. Partly it has been explained in Chapter one in context with "Dharm Kshetra". There is lot of misconception about DHARM. It is considered at par with religion. In SANSKRIT it is not the meaning as religion. Religion is a modern day, 18th Century concept as stated above. There is no definitive definition of religion. Oxford dictionary defines religion as belief in and worship of a superhuman controlling power, especially a personal god or gods. In English, the word religion has come to exist in recent times. Religion is a modern western concept. Divine, Sacred, Supernatural faith and belief is accepted as religion. The belief or faith is subject to change.

One may believe in something today and may believe in other thing later. Hinduism is considered "Sanatan Dharm". This is miss quoted being from Hindu text. Sanskrit literature has not mentioned Hindu word. This is the misquotation by the invaders of Bharatvarsh. But DHARM in Sanskrit is eternal to substance of manifest world, living or non-living. Pious duties and pure intellect for the welfare of human beings is the meaning of VEDIK scriptures about DHARM. Scientifically, maintaining structure and function of living objects and life forms, ultimate energy-spirit, living force is dharma of the living being or the manifest material world. It can be further explained by an example of fire. Light and heat are inherent in the word fire. Without light and heat fire has no meaning. This DHARM is eternal for the "fire". SANATAN DHARMA or "eternal religion". This is eternal to substance of moving or non-moving nature. (living or non-living is wrong hence not used here. as non-moving also have life. The whole universe is moving and living, has birth, growth, life span, change or growth, produce new ones and death). Details are in chapter 4 of this book and shall be discussed appropriately. Therefore, the word DHARM should not be considered at par with religion.

With religion other words need some discussion. Other words used are PANTH, Sect and SAMPRADAAY. PANTH is referred as laws that guide religious practice and faith. Faith is in demonstrable ideas but when it is in unknown or non-demonstrable, may it be 'God', it is BLIND FAITH. It develops gradually. While SAMPRADAAY is tradition, spiritual lineage, religious system that also forms SECT or CULT or tradition.

All religions are basically same. But the selfishness, greed, feel of supremacy over others force some to act and govern others to follow their vims and fancies for their ultimate benefits. This is responsible for hatred, enmity and causing torture to other fellow being, for obtaining material benefits from society, sect, faith, SAMPRADAY or culture. **Because of this many scientists in modern history have suffered insults, hatred, and even punishment in the hands of "custodians** of religion". Essentially that cannot be considered as religious. As said earlier, material involvement distances minds from spiritual benefits. One must strike the light of spirituality within oneself. The "Absolute truth" is everywhere. It is within everyone as well. Realize it. This is VEDIK teaching. Good for the society and the whole mankind, all nations of the world and for the progress of the science.

Persons who are led by material conception of life do not know that the aim of life is realization of ABSOLUTE TRUTH. As per Vedik teaching, the persons who do not know the liberation from material bondage are called as NON-ARYANS (highlighted and underlined part of stanza below). Contrary to this the word Aryan is applied to persons who know the value of life and have a civilization based on spiritual realization, striving for the knowledge of absolute truth. "Arya" is not a race or one of the tribes, it is the attitude and behaviour of person or persons. Every human who behaves and acts like this (as said above) is an ARYA, irrespective of the country, cast or creed etc. This is stated in Gita 2. 2. (page36)

Arjun was a KSHATRIYA and deviation from duties of Kshatriya i.e. not to fight war is against Kshatriya Dharm. This cowardice is akin to be a NON-ARYAN. Compassion to kinsmen may not lead to name or fame, on the contrary such acts point towards cowardice. Such acts may not be considered as magnanimity towards GURUS, GRANDFATHERS etc. but as an act of cowardice. It is a weakness of heart. It is against the DHARMA as KSHATRIYA. GITA and in fact VED have preached this to all human race and needs to be followed for the welfare of human society.

Fascinatingly in 15th to 20th century RELIGION has divided humanity under the pretext of faith, belief, cult, politics, Sampraday and technologies. With easier travel opportunities, the humans of different regions and continents have come closer with cultural pluralism, than before. Vedik Sanskaar are just the right kind of teaching to younger generations. This is needed more at the present time.

After imagining and visualizing mentally the post war consequences, war with teachers, kith and kinsmen, friends, relatives ARJUN was confused about what to do – "KIMKARTAVYA VIMUDH" or bewildered state of (Foolish and workless state of mind) mental situation, hence took refuse to SHRI KRISHNA. A "science of Krishna consciousness or super soul consciousness" is important here. What is science of Krishna or super soul consciousness? It needs some elaboration. From other sources like statements from PADMA PURAN etc. One who is master in, science of consciousness is a real spiritual Master. If a person is learned scholar in VEDIK wisdom (VIPRA विप्र), or born in lower family, or in renounced order of life but has mastered science of super soul consciousness, he is perfect and bona fide spiritual Master (Chaitanya Charitamrita). PADMA PURAN comments about this that, who can be a spiritual master:

Shatakarma nipuno vipro mantra - tantra visharadah I
avaishnavo gururn syad vaishnavah shvapacho guruh I

Scholarly expert BRAHMAN with expertise in subjects of VEDIK knowledge, may be unfit as spiritual master without mastery about science of super soul consciousness. But a person born in lower caste can be a spiritual master if he is expert in "science of soul Consciousness".

Following stanzas of GITA are in fact, pertain to recently developing science "SPIRITUAL SCIENCE" for the modern world. Centuries ago, this was highly developed in BHARATVARSH (later named as India with vested interest of invaders). Therefore, what Shri Krishna explained to ARJUN with examples which were prevalent in the contemporary society of BHARATVARSHA of "SANSKRIT SPEAKING CIVILIZATION". It is a common practice in teaching that a teacher cites examples which are well known in the society and public at large, for better understanding.

How funny was the situation in MAHABHARAT war! Rival enemy forces were cousins, relatives, teachers, friends etc. A fight between righteous and evil mentality. Teachers and students in opposite rival forces. Grand children and grandfathers in fighting mood facing each other in the field for war. Any sane person would think what ARJUN thought and so depressed to have made up his mind for not to fight as one cannot enjoy the post war success or failure after killing his own kith and kin or getting killed by them. (Even today, this Arjun tendency is present in the minds of "evil fearing" population. Situation was, exactly same as at the time of invasion by invaders in India. There was BUDHISM on top. Invader CHANGESE KHAN was from BUDHIST region – Mongolia. Same as that of India. Gita teaching is as relevant today as it was during Mahabharat period.). **Shri Krishna expert in VEDIK knowledge and wisdom, highly illumined, guided ARJUN with the help of SPIRITUAL SCIENCE.** SANKHYA PHILOSOPHY was prevalent in society. Shri Krishna took help of that to convince ARJUN. Before we understand details of SANKHYA, the scientific fact must be understood. *Modern science has started believing that BRAIN responds and adopts anatomically and physiologically to MEDITATIVE practices. Neuroscience research has revealed some surprising effects of spiritual practices on the brain.*

(Conscious Lifestyle magazine. Spring 2014 issue, Spiritual Science: New study reveals the surprising effects of spirituality on the brain by Justin FAERMAN – A new collaborative study between Professors of Columbia University and the New York state Psychiatric Institute).

Details of Science of KARM YOG has been described in 3rd chapter of this book. How it works has a scientific explanation. In transcendental meditation, SILENCE is important. "**Science of silence is SILENT DYNAMISM**". Meditation is basically a state of realization of self. In meditational awaken ness, intelligence is fully awakened (some people sleep during meditation – it is not meditation. A "Yogic Nidra" is a separate subject.). Consciousness is the basis of all creations and evolutions. **Infinite dynamism is stored in infinite silence. Same exists in self-realization in meditation.** To explain this scientifically, it is known that speed of motion in waves, is always slow because speed gets disturbed in ups, downs and sideward movements.

It consumes more energy. When movement is in a straight line and free from such ups and downs and at the same time free from interference or negligible interference, **in the straight and steady movement, speed is faster and energy consumption is minimum. It yields more profit.** Similarly, in transcendental meditation the mind is steady and straight line of active consciousness – in awakened consciousness in silence. For this act the time advocated is BRAHM MUHURT (a time 3 AM till before sun rise. This time is also scientifically selected) Mind gets full benefits of laws of nature with confidential and universal knowledge. Performance of action with establishing in YOGA i.e. unity with super soul abandoning success or failure is yoga unity. (*Yogasth kuru karmani Gita 2, 48*). Therefore, the yoga is the art of all work in material world in all walks of life (Yogah karmasu koushalam Gita 2, 50).

This probably is the basis of VEDIK knowledge. Understanding of such knowledge modern man can travel in universe with mental consciousness leaving material body by SUKSHMA SHARIR (VED and PATANJALI YOG SHASTRA have described three bodies of man – STHOOL body, SUKSHMA body and AATMA) It is the conscious body with which one can travel planets of solar system. This has been proved by Dr. VARTAK a medical professional, in his book – "UPNISHADANCHE VIDNYANNISHTH NIRUPAN by Dr. P.V. VARTAK, 7th edition, 2018. Page 8-16. To prove superiority of spiritual science he experimented with spiritual science. In 1975, USA was to send VIKING 1 to mars. After 11 months it was to return to earth. In transcendental meditation Dr. Vartak achieved SAMADHI and reached Mars planet with SUKSHMA SHARIR (Astral body). He published his observations in a Magazine "SANT KRIPA" on 21st May 1976 – (JUNE volume). After 2 months i.e. 21st July the messages started coming from Viking 1. Interestingly the observations of Dr. Vartak tallied with the observations sent by Viking. Thus, Dr. Vartak has proved his visit to mars under spiritual science. Leaving aside this experimental fact by Dr. Vartak, in **TAITARIYA UPNISHAD** IT IS

CLEARLY NARRATED THAT BODY CAN BE GROUPED INTO FIVE DIFFERENT SECTIONS. e.g.

1. ANNA-RAS-MAY (food body) basis of this is earth. Can move on earth.
2. PRANMAY (ethereal or air body) basis of this is VAYU-air. Can move in air upto limit of available air.
3. MANOMAY (mind body) basis of this is VAYU-air. Can move in air and beyond.
4. VIDNYANMAY (Science body) basis is beyond air MAHA, Can move beyond atmosphere.
5. ANANDMAY (Joy body) basis is spirit – BRAHM. Can move unlimited. All can be considered as micro bodies (SUKSHMA SHARIR) in Sanskrit it is—

Tasyaiv eva sharira Aatma I yah purvasya I tasmadava atasmadivdnyanamyat I anyontara aatmanandmayah I teneishapurnah sa va esha purushavidha eva I tasya tam I anvyam purushavidha tasya priyametra shirah I modo dakshinah pkshah I Pramod uttarah pakshah I Aanand aatma I brahma puchhya pratishtha I tadapyesh shloko bhavati II 2 II

(TAITARIYA UPNISHAD,
Brahmanand valli - 2, Anuvak 3, stanza 2).

The subtle types of bodies are described above and in later chapters of this book. '**Pran-may** Atma' is the soul of 'Ann ras may' (Body made from food) sharir, i.e. bodies described in above stanza. That Atma is "MANO-MAY". The movement and visit of micro body to distant planets being a testimony of this as stated above. Modern science needs a targeted research on this subject.

ARJUN argued that DHARMIK principles should be given more importance than politics or sociology. Science of KRISHNA consciousness made him believe that knowledge of MATTER, SOUL and SUPREME soul is even more important than religious thoughts and practices. A quote from ISHAVASYOPNISHAD can clear the dilemma of Arjun (or – reader. Because Gita is for the Global humanity), that spiritual and physical/worldly knowledge both are important in this material world:

Vidyam chavidyam cha yastadvedobhaya saha I
Avidyam mrityum tirtva vidyayamritamshnute
(ISHAVASYOPNISHAD. 11).

Spiritual knowledge is VIDYA. The spiritual and physical knowledge, both, are helpful in material world. With the help of physical knowledge, one wins over material problems and with the help of spiritual attains AMRIT TATVA OR BRAHM. The spiritual knowledge may not help person (with spiritual knowledge only), out of cancer disease problem or defend country from the wrath of enemies. To handle such problems physical knowledge is important like medical support and therapy and knowledge of warfare etc. **Over all view of India's (BHARATVARSH'S) past development, clearly depict the development of physical knowledge like Medical, physics, chemistry, Architect, Economics, Metallurgy, War instruments so on and so forth.** Every physical science was highly developed and used for physical benefits. Such things are not due to spiritual knowledge. The Spiritual knowledge was also on highest degree of material world. That is why the Sanskrit speaking civilization was highly developed and prosperous during that time. It was known as GOLDEN BIRD. The tremendous wealth that attracted invaders. Nonviolence, humanity, peace, neglect of defence system, resulted in slavery for centuries.

In this context Astrology ("Jyotish") and spiritualism (Adhyatm) is worth quoting. Rish-Muni and sages, who contributed a lot to Bhartiya culture (Sanskriti), divided Indian culture into Physical and spiritual. The Physical, material achievements for pleasure and peace several LOGIES (Shastra) were developed. Aayurved (medical science), Jyotish Shastra (Astrology, Astronomy), Yog Shastra (Yog and Meditation), Vastu Shastra (Architect), Shilp kala (Art and Craft), VAARTA Shastra (Meteorology). In life there are Favourable and unfavourable times.

Destiny depends upon Work and action Jyotish Shastra guides present and future life. Main purpose of Astrology is guide life to spiritualism. Nature's manifestation, protection and annihilation depends upon time cycle (KAAL CHAKRA/). Time cycle is Second, Minute, Hour, Day, night, months, seasons, Southern and Northern Sorties, year, Era (Yug), and Life span of Manu {MANVANTAR – There are 14 Manvantars (of Brahm) all described in later chapters at appropriate place }. Sun planets and satellites make time move and always mobile. In fact, Solar system is witness of living beings's work and actions.

Manifestation and annihilation of Nature goes on cyclically is called KAAL Chakra or Time cycle. Gita reiterates this in Chapter 8, Stanzas 16-19. Everything is under control of Time Cycle (details in Chapter 8).

One must raise oneself a little above the material world and think in a different way. How this material world has come to exist? WHERE FROM THE ENERGY is coming for the life in the material world? If nature has created this material world how the nature came to exist? The stars, planets, galaxies are responsible for nature, how stars etc. got space for these to accommodate? Science understands and recognizes if proof is available. Vedik knowledge has always pondered on this issue. Sanskrit literature is full of such subjects. Analytical discussions exist in Ved and Puran. There is some supreme which is responsible for all this. If one sees smoke, it is expected and imagined that fire is somewhere there, even one does not see fire directly. Therefore, the above statements indicate existence of supreme authority of the universe. All this comes to brain when one ponders on the subject in transcendental meditation. Knowledge is over all supreme in material world of this perishable universe. Hence Gita 4, 37 says:

Yatedhansi samiddhognirbhasmasat kuruterjun I
dnyanagnihi sarvakarmani bhasmasat kurute tatha II 37 II.
(Gita 4, 37).

Firewood burns to ashes due to fire so the knowledge burns reactions and ignorance of reactions of material activities.

The discussion of chapter 2 of Gita indicate that the physical material body is essential for the soul to be a part of the body. The micro bodies or subtle bodies dwell into gross material body. The senses are the part of the physical body. Mind and intellect are outcome of the physical material body. All this is essential for the cultivation and piousness for the liberation. The body is instrument for the purpose of liberation of soul. How the bodies are formed. What makes the living beings to get manifested in this universe. Sanskrit literature dealt this subject on basic and root cause of Brahmotpatti – manifestation of universe before the Prajotpatti – manifestation of living beings.

Interestingly in probing the knowledge regarding PRAJOTPATTI – how population came into existence a question is asked in PRASHNOPNISHAD:

Atha kabandhi katyayana upetya paprachha I
bhagvankuto ha va imah prajah prajayant iti II3II
(Question 3 of Prashnopnishad - of Atharva ved).

PRAJA has a wide meaning in Sanskrit literature. It means all living beings on earth not only humans. Modern science has till date found no answer or modern science is silent on this. But the question has been answered in VED in the form of answer to the question:

Tasmei sa hovach prajakamo vei Prajapatihi sa tapotapyat sa
tapastaptva sa mithunmutpadyate I
rayim ch pranam chetyeto me bhudha prajah krishyat iti II 4 II
(question 4 of Prashnopnishad)

The answer goes beyond sex and embryology. A pre-embryology stage of PRAJOTPATTI. Creation of all living beings on earth. Creation of RAYI and PRAN (details are in later Chapters of this book). Brahma was alone in the beginning. He desired to create living entities. It is sufficient here, to mention that the CREATOR struggled with perseverance, (and bodily mortification) resulting into fire or heat. HE the creator manifested a pair (Utpadyate) **not male and female** but

RAYI and PRAN. Rayi is matter and PRAN is energy. Brahm divided itself into two (pair) – SHIV and SHAKTI. Both are not having any form or virtue. From this, the whole living beings manifested in this universe.

Perplexed and confused Arjun did not ask for spiritual knowledge from Shri Krishna. He simply projected social concepts and traditions prevalent at Mahabharat time, but Lord Krishna suddenly started the SANKHYA philosophy about soul and super soul. Arjun overpowered by traditional psychological thinking. Arjun talking like a learned man, but a learned man knows that what is body and what is soul. Whatever is born, must end is law of universe. Therefore, the body born, must end someday.

Ashochyananvashochastvam pradnyavadanshcha bhashse I
gatasungatasunshcha nanushochanti panditah IIII II

(Gita 2, 11).

Arjun talks like a knowledgeable man but lamenting on not worthy of it. Persons come (birth) and go (death) but wise with wisdom, do not lament on the same.

It is the soul or energy which runs the show in a body. Once soul (energy) leaves, the body is non-living. So, the soul is important and not the body. Hence there is no cause for lamentation for the learned person. Existence is an incessant process in this universe. All these people present in the field were existing before and will be in future also (see above). No need to grief for material body. It is an eternal fact which is confirmed by VED and various ACHARYAS. The living entity is an individual soul. Living body is changing every moment. Manifestation of body as Child, Youth, Old and so on. But the spirit remains same unchanged. At death spirit migrates to another body. Understanding such a knowledge in a man is known as sober personality – DHEER. It is like happiness and distress which disappear like seasons in environment. It is only a sense perception. Tolerance, behaviour without disturbance is sanity. Proper discharge of duty by a person, without any disturbance. Disturbances are trivial and uncalled for. Person who is steady in adverse situations, is a DHEER person. Certainly, he is fit for liberation.

The body changes by action and reactions of cells present in the body. It is observation of medical science. But spirit, soul or energy present in the living body remains uninfluenced in all stages of the body. In other words, the body changes but soul is steady and eternal. This is the difference between the material body and the spirit. Bewilderment by the ignorant influences must be shun by a learned is a teaching of dialogs between ARJUN and KRISHNA in GITA. To drive away the ignorance, BHAGWATGITA has been taught by Shri Krishna, for the enlightenment of all humans for all time. In this context Lord Krishna has warned in 7th chapter stanza 20th that out of fear (or vested interest), persons whose intelligence is stolen by material desires, surrender to DEMI GODS. Such persons are nonspiritual as per their nature. Such people tend to forget the Supreme soul:

Kameisteterhrutdnyanah prapadyntenyadevtah I
tam tam niymamasthay prakritya niyatah swaya II20 II
(Gita 7. 20).

Scientific explanation of ATMA in all living and nonliving objects of this universe in the stanzas of GITA and can be authenticated further from the quotations of UPNISHADS –

(mandukopnishad, kathopnishad and shwetashwataropnishad – stanzas quoted in original).

ATMA the soul which is in living being is indestructible and imperishable.

Balagrashatabhagasya shatadha kalpitasya cha I
bhago jeevah sa vidnyeiyah sa chananatyay kalpte
keshagrashatbhagasya shatanshah sadrashatmkah I
jeevah sukshmaswarupoyam sankhyatito hi chitkanah II 9 II
(Shwetashvatara upanishad 5, 9).

Ashorunatma chetasa veditavyo sminpranah panchadha samvivesha I
praneshchittam sarvamotam prajanam yasminvishudhe vibhavatyesha aatma II 9 II
(Mundakopnishad 3, 1, 9).

Tip of hair, split into 100 parts such a small fraction of soul is in living body Even a common man can realize that when spirit/soul leaves body, the physical body is lifeless i.e. body is dead. A small fragment of the supreme spirit is soul. The soul, an atomic fragment is in fact a part of the supreme soul. The size is so small that it cannot be measured. It is obvious that the body is maintained by spirit/soul. Similar stanza has appeared in KATHOPNISHAD (1.2.18). It is exactly like GITA SHLOK: (compare stanza 20 of Gita chapter 2).

Na jayate mriyate va vipashchinnayam kutahshchinna babhuva kashchit I
ajo nityah shashwatoyam purano na hanyate hanyamane sharirellI18II
(Kathopnishad 1, 2, 18).

The SOUL – a small part of the energy of whole UNIVERSE which is fully aware and knowledgeable or in other words the soul is always aware of knowledge. But the difference is the PARAMCHETANA which is fully knowledgeable while the soul CHETANA is limitedly knowledgeable. As it is limited to the body in which it is trapped. This CHETANA forgets its source i.e. the trapped soul/energy forgets its original being – PARAMATMA. In MUNDAKOPNISHAD 1. 1 and 2 stanzas it is said:

Sa brahmavidyam sarva vidyapratishthita I
matharvayjyeshthaputraya prah II 2 II,
(MUNDAKOPNISHAD 1. 1 and 2 stanzas).

Spiritual science is the source of all other sciences. How to know the soul or BRAHM has been described in the stanza of MUNDAKOPNISHAD: quoted above on page 42. (MUNDAKOPNISHAD. chapter 3. 1. 9.)

The ATMA or subtle self can be realized by MIND or consciousness, but consciousness need be cleared, as it is entangled in sense enjoyment, sense gratification, then the self comes to light by itself. It is not attained by discourses, nor through intellect, nor through much hearing about it. It is attained by oneself, alone who longs for it whole heartedly. The self of such a person reveals its nature. This simple and secret knowledge is narrated in GITA. The rivers merge into ocean leaving name and form similarly the illuminated man with realized soul, finds its way to the "Supreme Being". It reaches higher than the camouflaged world (MAYA-माया).

Anoraniyanmahato mahiyanatmasya jantornihito guhayam I
tamakratuh pashyati veetshoko dhatuh prasadanmahimanmatmnah
(-KATHOPNISHAD 1.2.20)

Here it is important to understand that the PARAM ATMA original, and body ATMA both dwell in the body of the living being be it be plant or insect/animal etc. the one who has understood and given up all desires and all living concerns only can understand the secret meaning of ATMA and PARAMATMA.

The individual soul is everlasting and eternal, present everywhere unchangeable immovable and basically eternally the same. While performing bodily duty if body gets killed or assassinated by other, in war, both enemy forces are free from lamentation after understanding glories of the LORD.

This is what Shri KRISHNA explaining in GITA to ARJUN. When Arjun understands and realizes this, from guru, friend Shri Krishna, the desires of ARJUN, completely disappear and he becomes free of all bondages.

Eternal existence of soul is claimed here in the stanza. The soul in living things in the universe around us includes Water, Air, Earth, fire and space (आकाश) i.e. living entities all over are GOD'S CREATIONS. This includes fire as well. Common belief is that the fire sterilizes the material under fire. The following stanza is contrary to this belief, as far as the soul is concerned:

Achhedyo yamadahyo yamakledyo shoshya eiv cha I
nityah sarvagatah sthanurchloyam sanatanah II 24 II

(Gita 2, 24).

The soul cannot be burned by fire. In other words, the living entities are also present on the star SUN. Forms of such living beings may be different. Suitable bodies provided to living entities to live there on sun. Sun has energy from supreme soul. If this is not possible then the word of SANSKRIT **"nityasarvagatah", the meaning of this i.e.** living everywhere is meaningless. Science may not agree to this. **But the statement is according to scientific norms cannot be denied. It is as per the laws of constitution of nature. The laws of nature are intricate but uniformly applicable all over in universe (see general considerations of this book).** Living bodies in universe are revolving between manifest and un-manifest i.e. physical and chemical combinations of living objects are un-manifest in the beginning, manifest for some time and after death un-manifest again. Hence why one should lament on such temporary issues. According to "anoraniyanmahatomahiyan" theory gigantic animals like Elephants and micro – organisms (plants like Banyan tree and micro Viruses/bacteria), in nature have atomic soul. Persons engrossed in SUBJECT MATTERS (माया), it is difficult to understand the theory or the facts of soul and PARAM Aatma or supreme SOUL.

Avyakto yamachintyo yamavikaryo yamuchyate I
tasmadevam viditvenam nanushochanti panditah II 25 II

(Gita 2. 25).

The soul is unmanifest, invisible, unchangeable, unconceivable, knowing this Arjun need not lament.

Avyaktadini bhutani vyakta madhyani bharat I
Avyaktanidhnanyev tatra ka parivednah II28II

(Gita 2, 28)

A poem worth quoting here.

titled: Vyakta avyakta:

Kuni kunache nahi jagati sarvachi apule Jeevan jagati
Paradheena jagi sarvachi asati deiv davil tikade jati II 1 II
Jeevatuni jeevachi nighati kamananduni visarun jati
Kalakar anabhgya asata lapuni teer laksha vedhata II 2 II
Tyagun dhanur teersarasavati teerachi pudhache teerutpadati
Kale kuna na kase he ghadati mayajali yukti bhalati II 3 II
Yug yugantar asechi jati vyakta avyakta houni jati
Bhut kadhina vykta mhanavati vyakta Madhya sarv janati II 4 II
Techy samsara chalavati na kalat bhavishya chalun yeti
Bhut kunas na athavati bhavishyat koni javu na shakati II 5 II
Na kalat jag he vahat asati Ishwar kunas kale na jagati
Mayecha khel ha sara mayetchi sare guraphatati
Adnyanate nighuna shakati sant dnyani ishawar aathawati II 6 II

('satyabrahma jaganmitthya' B G. Matapurkar,
publishar: Nandini Santosh
Tamboli. ISBN: 978-81-907405-1-7).

"Manifest Unmanifest"

No body belongs to any body. All live their own life. All are dependent on fate and behave as directed. Living beings are born through living ones and forget after enjoyment. Artist of universe is unaware of this. Invisible and aiming arrows to targets. The arrows leaving bows and arrows create arrows of future. How this happens no one knows. By trick camouflage the nature. Time passes away and manifest become unmanifest. Past is never manifests. All understand only manifest who run the show. Without knowledge future reveals. No one knows past but cant visit future. All world goes like this. God remain unrevealed. The game of camouflaged world continues and no one can come out of this. Knowledgeable persons only know GOD.

Modern science is developing stage by stage, and it is still incomplete and in developing stage. Matter, substance, and molecule are the most major concepts in Physics, Chemistry. Most knowledge about matter derived from Greek philosopher Aristotle's theory of "Matter" which is made up of 4 elements – Water, Air, Fire, and earth (384-322 BCE). DEMOCRATUS (460-370 BCE) suggested divisibility of matter. Atom being the smallest particle of matter. It is indivisible. Air, Water, Earth etc. are all consists of small particles. In the beginning of 20th century W. Ostwald was against the atomic theory and suggested that the **matter is nothing but a complex of energy factor.** (See MAHARISHI KANAAD. Explained in chapter 3 of this book). It will be sufficient here to mention that:

According to MAHARISHI KANAAD, (7000 B.C.) THE 8 MATERIAL TYPES OF NATURE, are all considered matter and of the entire universe. Earth is solid matter, water is liquid matter, Air in gaseous form of matter. He considered even the brilliance as matter. He goes further to say space and time are also matter. (Vaisheshik Darshan 1. 1. Stanza 4 and 5). Modern science, now started considering matter and energy as one (W. Ostwald) – see highlighted part of stanza below:

Pruthivyapstejovayurakasham kalo digatma
mana iti dravyani II 5 II
(Vaisheshik Darshan 1. 1. 5).

On splitting the underlined, highlighted words one gets: Earth (PRITHVI), water (AAP), Fire (TEJ), Air (VAYU), Space (AAKASH), Time (KAAL), Direction (Disha), Soul (AATMA), mind (MAN), ITI DRAVYANI (all are DRAVYA = PADARTH=substance). Space, sky or AAKASH, is also matter which is without atom. Aakash being a matter has no atom or has no particle. This is for the facilitation of movements in space. Therefore, all movement take place in Space or Aakash, which is **influence zone.**

In the form of Shri MADBHAGWAT GITA, Shri Krishn while explaining to ARJUN with examples, which were commonly prevalent in contemporary society of that time. The energy factor or soul factor, as in above stanza – is present in all the material world moving or non-moving substances. This suggests the highest degree of scientific development and understanding of common people. In stanzas 22-25, capacity of soul is narrated. It is unborn, eternal (stanza 20), inexhaustible, formless (Stanza 21). It is capable of adopting, new and different bodies (stanza 22). It is in water but cannot be moistened. It is in fire but cannot be burned. It is in air but cannot be dried, and so on. Scientifically speaking soul is present in different form and bodies and bodies adopts the local surroundings, and capable of survival. The living forms are present in extreme heat and cold atmospheres.

Vedik knowledge speaks so many things about soul and super soul. But what is the scientific evidence about existence of soul? Ved philosophy believes in immortality of soul while Abrahamic or semitic religion/faith, and dialectical materialistic societies do not believe in immortality of soul. Believe in material nature. But there is no answer to the question how the nature came into existence. As stated earlier, that the presence of AURA around all living entities is to some extent confirmed by KIRILIAN images. The aura is emanation of light energy of soul or physical energy (SHAKTI). The mystic light is acceptable by quantum theory. Quantum of radiated light energy gives idea about electron movement. The soul energy remains even after death of material body. Stanza 13 of Gita says the same thing:

Tatha dehantara praptirdhirastatra na muhyati II 13 II

Changes of physical body child adult old age etc but no change of the energy or soul in the body. After death JEEVVATMA or soul adopts another physical body. Hence is eternal.

Science of KARM YOG or work action: After keeping free from bondages of body and ignorance, Shri Krishna explains how to keep away from bondages of work and senses and its results involved in the work and sense gratification. Arjun's sense gratification is responsible for the decision not to fight, and with emotionally engrossed, unknowingly sacrificing wisdom and duty. Work with intention to achieve fruits of work, is different from work with sense of duty and no intention of results i.e. good or bad whatever may it be, in the matter of sense gratification or material happiness. It is a transcendental quality of work. This makes one to be free from the tensions of the result outcome – good or bad. One can understand the SCIENCE involved in this attitude of work. Work is YADNYA in Sanskrit literature. Desired work or action is Yadnya. While action for big aim or big public utility is Maha Yadnya like ASHWAMEDH Yadnya. In Gita 3rd chapter stanza 9, 10, 11, and 12 it has been stressed to do work and actions to please demigods (*Bhumiraponalonilonabha:* Earth, Water, Fire, Air and Aakash) so that mutually both be benefited. After CREATING LIVING BEINGS i.e. PRAJOTPATTI Brahma announced to living beings or praja that by YADNYA KARM you achieve desired goal in your life (3. 11-12).

Devanbhavayatanen te deva bhavyantu va I
paraspara bhavayanta Shreya paravapsyatha II 11 II
(Gita 3, 11).

In the stanza it is clearly indicated that make gods rich by emotional approach in return get benefited thus by mutual help both will be benefited. Commonly the meaning is taken as by pleasing gods somewhere in "heaven" and in return get benefited. Contrarily it is otherwise. Good actions are important. Lord Krishn indicates here emotional centres within our material body. Stimulation of such centres with good actions would bless in return. It is known that GOD and DEMON are within us. The stress is on work. Mutually i.e. emotionally and materially it is beneficial. In next stanza it is further clarified that, powerful good actions would fetch good outcome:

Ishtanbhoganhi vo deva dasyante ydnyabhavitah I
(Gita 3, 12).

Emotional involvement in good work reap good benefits, is well known. Stress is again on YADNYA BHAV. Gita further clarifies in stanza 13 that, food stuff obtained by good work actions (YADNYA KARM) bears no stigma of sin. It is scientifically explained in next stanza:

Annadbhavati bhutani prjanyat anna sambhavah I
ydnyat bhavati parjanyo ydnya karma samudbhavah II14 II
                                              (Gita 3, 14).

All living beings are due to food. The food is the result of rain and environment (rain is par + janya i.e. depends upon something else, born due to other substances and events i.e. heat (sun, cosmic rays etc.) + water (oceans, sea etc.)= steam = cloud, then rain). All this is possible due to good actions. Protection of atmosphere is mutually beneficial, needs no explanation. Scientific analysis is still deep. ANNA does not mean only "Food" but all substances essential for life, which keep life pulsating or vibrating. It involves whole cosmos and essential cosmic substances e.g. PRAN, essential gasses etc. This involves whole cosmos, from atom to all living including human beings (BHUT). PARJANYA means not only rain but essential energy or PRAN for life to vibrate. The whole lyrics in the stanza is scientific only if understood in its real perspective.

Physical work and its benefits end with end of physical body but work with transcendental work is ever lasting and even carried forward for next incarnation as well. So, no loss at all, of any kind. What a great psycho-analytical philosophy is this, which is highly scientific. One feels no stress of work or its result. Achieving a great solace at the end. In material world one is hurt when work is incomplete and happy when work is complete. But yogi is least worried for the work completion. Even if the work is not done, he is happy because he has devoted and dedicated the work to GOD. In next attempt it will be completed. No feeling of sorry. He is not attached to results of the work. Happiness or sorrow or success and failure profit or loss all are equal for the yogi or transcendentally conscious person. Mentally he fixes all to the LORD. Keeping himself free from all bondages of material world.

Equality in the success or failure of work and abandoning all attachments into success or failure is considered as YOG. This is devotional service. Shri Krishna proposes to Arjun that follow YOG and strive for YOG. That is the art of all the work. He is an advocator of war in the war field and fight. Kill people assembled on the war field. It is violence against people. The circumstances are important here. The rival forces in war field, were detrimental at large to all humans. The war erupted between greedy and evil forces against the just, caring for social norms and for the good of humanity which is MANAV DHARM:

Svadharmapi cha vekshya na vikampitumarhasi I
dharmyadhi yudha chchheyonyatkshatriyasya na vidyate II 31 II
(Gita 2, 31).

In other words, God's order justifies war killing but otherwise willingly killing cannot be justified. When war is forced upon somebody then as KSHATRIYA, one need not hesitate to war killing. May this killing be one's own relatives having greedy and evil intensions detrimental to society. Question is who will decide this because both sides have their own arguments in favour and against.

Selfish motives of the KAURAV group supported by the ruling King and the supporters (considering themselves as duty bound) including teachers, friends and grandfather etc. all forces were bent upon destroying social structure of the country. Every nuke and corner to establish the peace and harmony failed due to adamancy and emotionlessness of DURYODHAN, the acting King. Under the circumstances there was no option left for compromise. The situation was same as during Hitler's time and Japan's attack in "second world war". Despite the logical conclusion on compassionate ground, of ARJUN's decision not to fight, Shri KRISHNA labelled it cowardice and anti-KSHATRIYA DHARM. It is left to reader to decide that violence was for supreme justice or not. Even today the hanging is ordered by 'Supreme court', a violence against the criminals for the administration of justice. Surgery is for the betterment of the suffering of patient. It cannot be considered a sinful action. This is apart from spiritual context preached by GITA.

VEDANT philosophy has always propagated the idea of four things in life –

1. DHARMA (not religion but basic quality, virtue/duty),
2. ARTH (Earning needful necessities),
3. KAM (progeny for next generation), and
4. MOKSHA (Liberation).

These are the four life's aims to be followed. After completing first three aims and fulfilling worldly desires, liberation is also important. This is the aim of life for human societies of all world even today and needs to be followed. Ultimately living life to fullest extent was the aim of Sanskrit speaking civilization of VEDIK period. Use of wisdom of age and experience, work for the good for the next generation.

CHAPTER THREE

BEHAVIORAL SCIENCE – SCIENCE OF KARMAYOG SCIENCE OF DEATH AND REINCARNATION

(KARM ORIENTED)

Abstract: After Buddhi yog the KARM yog is explained. Again, Lord Krishn addressing Arjun, explains the whole humanity, that the difference between the knowledge (DNYAN YOG) and the deeds (KARM YOG), at the same time what is beneficial to all humans. The action after knowledge without any expectation of fruit, is important for YOG of renunciation or wisdom. It is advocated by writer of Gita that work, action, deed performed with total surrender to super soul will take the person to supreme soul.

Vedik philosophy is based on the principle of looking deep inside and least external but keeping external surrounding in mind at the same time. Introspection and knowledge of "inner-self" is foremost. Here the mind is of utmost importance hence be under "self-control" lest mind will control you. After understanding the intentions of Lord Krishna, thus so far, ARJUN counter questions that why KRISHN want, ARJUN to undertake the horrifying warfare with all his relatives,

Guru, and friends. If the intelligence and Transcendental consciousness is better than the work, then why fight? Probably, this is a mindset of a science student. Lord Krishna consciously diverts attention of Arjun from fruitless work of transcendental conscious mind to KARM YOG. But Arjun demands that his mind experiencing a bias, after listening to KRISHN'S multi meaning suggestions, and requests to explain and suggest what is good and beneficial for taking decision. Bhagwan Krishna explained in this chapter to ARJUN and left decision making to Arjun. How scientific and what science is hidden in the stanzas one can decide from the following explanation.

Lord Shri Krishn in following stanzas has stressed that in the past, in this material world, two methods are described, DNYAN YOG, (SANKHYA YOG), and KARMA YOG – linking with devotion. This has been explained to ARJUN who is sinless (अनघ) to remove mental bias see stanza of Gita 3, 3 below:

Lokesmin dvividha pura pokta nishtha mayanagha I
dnyanyogen sankhyanam karmayogen yoginam II 3 II

(Gita. 3, 3).

Self-realization is the ultimate, aim of human incarnation in nature's material life. This is well decided opinion of RISHI, MUNI and Sages of SANSKRIT speaking civilization of VEDIK PERIOD. This is a fact which is difficult to accept if life is engrossed in material life. To understand it, many ways have been explained by SANKHYA YOG, KARMA YOG, BHAKTI YOG, BUDHI YOG, GYAN YOG etc. By asking such a question Arjun desires clarity for all students of human race to understand the mystery of Shri MADBHAGWAT GITA.

Self-realization can be achieved by empirical, philosophical speculation (BUDHI YOG). The soul is invisible to eyes, cannot be held by speech, nor be grasped by senses, unavailable by penance or action, but by wisdom and knowledge with pious mind set in transcendental meditation it can be experienced. Mundakopnishad says:

Na chakshusha guhyate napi vacha nanyeideveistapasa karmana va I
dnyanprasaden vishuddhastvastatastu tam pshyate
nishkalandhyaymanah II 8 II

(Mundakopnishad 3. I. 8).

Realization by devotional attitude of mind is BHAKTI YOG. Analytical study of NATURE of spirit-spiritual science (ATMA) and matter (MAYA) is SANKHYA YOG. Adopting BUDHI YOG all senses can be controlled. Person develops steady intelligence by keeping all senses under control by his thought process and mind. Pride and greed must be controlled to achieve this. This has been mentioned in Gita chapter 2, stanza 61:

Tani sarvani samyamya yukta aasit matparah I
vashe hi yasyendriyani tasya pradnya pratishthita II 61 II

(GITA 2, 61).

Devotional method is direct method and spiritual method is indirect where purification of senses is needed, to understand the self (ATMA) and super self (PARAMATMA) relationship.

It is a common belief in modern population in general that GITA preaches to renounce everything to achieve spiritualism. In next chapter stanza 4 it is clearly mentioned that success in SAMADHI and liberation cannot be achieved by simply renunciation. Mind and heart must be purified. Every person has to work or bound down to work, as per the quality, provided by destiny, or provided by material nature. But unfortunately, it has been ignored. Thus, Gita says that:

Na karmanamanarambhan neishkarmya purushoshnute I
na cha sanyasanadev sidhim samadhigachhyati II 4 II

(Gita 3, 4).

Renouncing work and duties assigned by nature, in the beginning nor adopting abandonment of total material world or abruptly adopting renunciation (SANYAS) one cannot achieve the final goal of SAMADHI.

There are four ASHRAM (life descriptions) which are to be enjoyed during life:

BRHMACHARYA ASHRAM (Study period,), GRIHASTHASHRAM (family life), VANPRASTHASHRAM (space for younger generation, Forest life) and SANSYASHRAM (Life without any attachment to material world). Purification by discharge of duties assigned or prescribed, by natural DHARM. This purifies hearts of people who are influenced by materialism. Material influence on mind and brain is natural due to effect of MAYA i.e. camouflaging effect of material world. Hence the VEDIK statement in order is DHARM (duty), ARTH (wealth), KAM (progeny) and MOKSHA (spiritualism and liberation), to achieve true liberation. **After sense gratification the detachment and liberation become easy.** MAHAMRITYUNJAY MANTRA states the same:

Om tryambakam yajamahe sugandhim Pushti vardhanam I
urvaruk bandhanat mrutyormokshiya mamrutat I
(YAJURVED).

(Also known as "Rudra mantra, trambyakam mantra)...

Shiv with three eyes, fragrant and virtuous, prayed and worshiped for liberation, who is provider of growth, nourishment, perfection and wealth to all living beings in material world. As the melon fruit once ripened spreads fragrance, drops out from its vine stem and gets liberated. Oh lord Liberate us.

Without soul or life energy the body is dead. Incapable of doing anything. It is the nature of soul to be active all the time. Even for a fraction of time soul entrapped in material body, cannot refrain from work. To engage it in good work or futile work depends upon the possessor of the soul. In contact with material energy the soul can adopt material modes of work. In order to purify the soul, must be engaged in prescribed duties which are good for the person. A warning is essential here that if a person controls his senses of work but mentally involved in thinking for gratification of senses it is of no use and he is simply a pretender (MITHYACHARY). In this 3rd chapter of GITA, KARM YOG is emphasized. **One cannot remain without KARM even for a fraction of a second. Because nature has binding on living entities regarding KARMA bondage (कर्म बंधन). One must perform his assigned KARM, duty.**

Virtues and work action principle decides the VARNA division system of society. Which is established by supreme soul (Gita 4, 13). Karm is important. It must be understood properly. Not only this but, action – Karm, inaction – Akarm, and forbidden – NISHIDHA karm, be understood properly. Actions are very hard to understand due to its intricacies. Work is intensively progressive. Gita 4, 17 says that:

Karmanohyapi boudhavyayam boudhavyam cha vikarmanah I
akarmanashva boudhavyam gahana karmano gatihi II17 II

(Gita 4, 17).

Intricacies of action are hard to understand. Hence one must analyse action, inaction and forbidden action. With intellect, intellectual sees action in inaction and inaction in action, even when he is engaged in his normal activities of life:

Karmanyakarma yah pashyedkarmani cha karma yah I
sa budhimanmanushyeshu sa yuktah krutsnakarmakruta II 18 II

(Gita 4, 18).

According to the time cycle the KARM can be grouped into:

1. KRIYAMAAN KARM-work performed during present time in life.
2. SANCHIT KARM-work performed in the past.
3. PRARABDHA KARM-work performed during past incanations.

1. Kriyamaan Karm: is work performed in present time. As stated above during whole life person is not without work even for a fraction of second. Work performed during present time i.e. during life. In other words, the work performed during lifetime is Kriyamaan

Karm. A person is inclined to do work as per the pious or evil effect on the mind set (CHITTA). But the person is free to decide which way he should work. Firm decision on whatever the action he chooses, he is free. The strength of his mind set is important. He is free and has no bondage of any kind. His action is all his final decision irrespective of benefits offered.

2. Sanchit Karm: work performed in past. Systematically collected work. Deposits of work and action is always in plenty and compulsorily enjoy or suffer. But in Gita Lord Krishn says:

Yatheidhansi samiddhognirbhasmasat kuruterjuna I
dnyanagnihi sarvakarmani bhasmasat kurute tatha II 37 II.
(Gita 4, 37).

Fire of "Knowledge acts like fire and burns all results of actions.

3. Prarabhada Karm: Work performed during past incarnations. Person bears the results of his deeds or karm in past incarnations for which he is born. He will have to enjoy or suffer as per his karm results of previous births. He has no choice. The suffering or enjoyment depends upon one's own work and action in past incarnation. There is no reason then why to grumble. Accept it with firm mind set. It is important here to understand that engagement in pious activity in present life, will certainly secure happy life in next incarnations.

Concept of "Reincarnation" theory is used repeatedly during discussion on science of Karm. It is discussed in later chapters. Some explanation is essential here. For this let us refer to chapters 5 and 15 of Gita. As it has been mentioned in general considerations and in my research discussions (in published science journals), the process of material body formation is an automated process. It needs no direction from any source. On the contrary the fertilized ovum directs the body of mother to act and develop as per direction of the developing foetus. (Fertilized ovum kept in uterus of a 60 years female where the reproduction process has come to a standstill, the ovum forces the

60 YEAR female BODY TO DEVELOP AND BEHAVE TO THE TUNE OF FERTILIZED OVUM.) The ovum and the sperm are living cells and form physical and material body, but spiritual body is dependent upon the transmigrating soul. This soul carries the resultant work actions and material body grows into an independent personality of the future. Interestingly Gita in stanza 7 of chapter 15 speaks about the departing soul at the time of death:

Mameivansho jeevloke jeevbhutah sanatanah I
manah shashthanidriyani prakritisthani karshati II 7 II
(Gita 15, 7).

The entrapped soul is fraction of supreme soul. At death, the entrapped soul is liberated from material body. It enters spiritual body. It is also eternal like supreme soul. At death, the soul carries the created consciousness to next body or transmigrate to another body with all 5 senses and the mind. If the consciousness is fixed to lower category then it gets lower category body (cat, dog etc). If fixed to good, Godly category he gets demigod body or supreme soul. Common concept is after death, all is finished. In-reality, it is a new beginning. Depending upon Karm in previous body, new body is assigned. The soul struggles for existence. Consciousness is seed for cultivation of next body. In Mother's womb ovum and sperm are living cells capable of forming physical body. But soul cultivates the personality of the body, in mother's womb. The future behaviour of body depends upon soul and previous Karm of soul in last entrapment. The 5 senses and mind carried with soul govern the outcome. Karm oriented principle, "Karm sidhant" is involved in this theory.

Na kartrutvam na karmani lokasya srujati prabhu I
na karmaphalasanyogam swabhavastu pravartate II14 II
(Gita 5, 14).

Action reaction and deeds are social routine behaviour. There is no influence of god or supreme soul. The whole behaviour is dependent on deeds. The deeds of last birth are responsible for our present and future life. Soul brings quality of life to the transmigrated body. Fertilization is life event. No new life is created as life is already there in ovum and sperm as stated above and in the stanza 14 of chapter 5. Body develops in womb and the soul brings his own past cultivations through the body developing in mother's womb.

But the KARM must be without any attachment for result outcome (NISHKAM KARM). Desire behind the KARM is important. Motive behind KARM should not be selfish. The motive if not selfish or it is for ultimate good for humanity and all living beings, then under such circumstances the bad effect gets dissolved. To explain, the killing during war is not murder and has no bad effect but otherwise it is murder and will have bad effect. Arjun could not follow this intention of Lord KRISHN, hence the question arose to his mind. Abandonment of nature bound work (KARM) is treacherous. Therefore, do duty assigned by nature. Abandoning sense satisfaction at the same time the mind remains attached to those senses is bad. Forcefully suppressing actions or work – KARM and senses but mental attachment to those senses and sense satisfaction is fraudulent and hypocritical. Therefore control your thoughts which are forcing you not to perform your duty as a KSHATRIYA. Such thoughts would affect one's future. And change results of KARM. In fact, thoughts are seeds of KARM. Hence KARM will affect destiny. One cannot change one's destiny but can control KARM willfully by controlling senses thoughtfully. **One cannot change one's destiny but by spiritual awakening one can, definitely change the way one reacts to the effect of affecting destiny. Evolution of self and evolution of social "mind set" (of society spiritually, by transcendental meditation.**

How scientific is this explanation, for ARJUN to take decision to fight or not. This was the attempt of Sanskrit speaking civilization of BHARATVARSHA (INDIA). Unfortunate part of slavery is, even the name of the country has been changed by invaders and the rulers. Nowhere in the world, the name of a dominated country, has been changed. This is in fact is a calculated attempt by the rulers. The Sanskrit speaking civilization was highly educated and scientifically oriented with 100% literacy rate, no one was beggar or poor, no one was thief. Science was highly developed. Testimony to this is the statement of Lord Macaulay in British parliament during early 1800. Newspaper cutting is reproduced here, words in the illustration are self-explanatory:

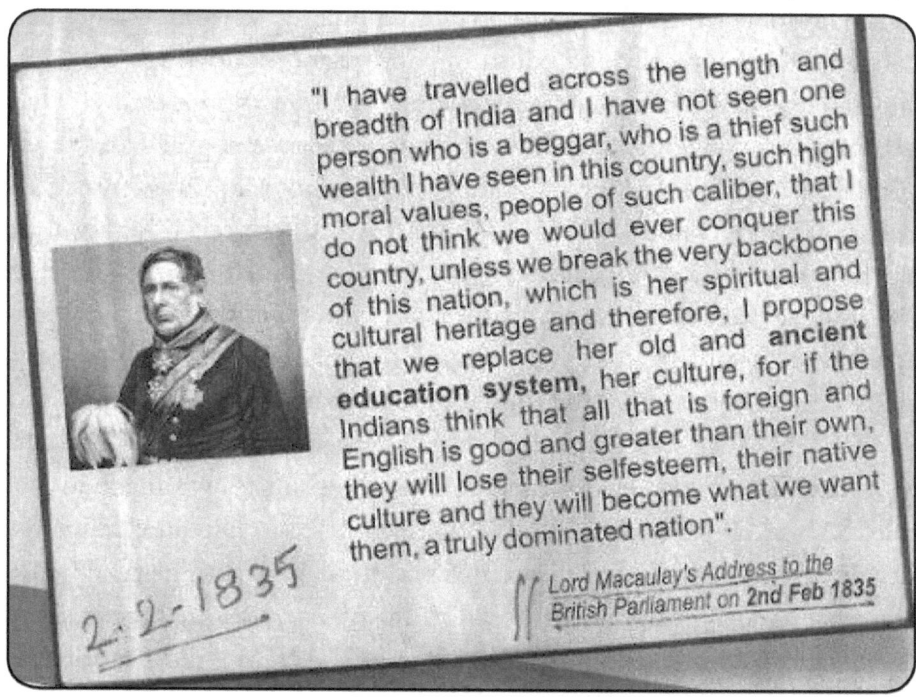

"I have travelled across the length and breadth of India and I have not seen one person who is a beggar, who is a thief such wealth I have seen in this country, such high moral values, people of such caliber, that I do not think we would ever conquer this country, unless we break the very backbone of this nation, which is her spiritual and cultural heritage and therefore, I propose that we replace her old and **ancient education system,** her culture, for if the Indians think that all that is foreign and English is good and greater than their own, they will lose their selfesteem, their native culture and they will become what we want them, a truly dominated nation".

Lord Macaulay's Address to the British Parliament on 2nd Feb 1835

2-2-1835

Cultural and spiritual heritage of BHARATVARSHA which was backbone of the country was broken. It was a calculated assassination of the spiritual, cultural and scientific mind set of the country. **It is important to note that in this UNIVERSE nothing is permanent. Whatever begins, must come to an end someday. Many civilizations have vanished from the earth. Under this universal law. Sanskrit speaking civilization has almost vanished.** But its roots are so strong because it is eternal (SANATAN) that it has started reviving. It is taking time. That is why the eternal DHARM established by that civilization is deep rooted and perpetual. Hence it is called SANATAN DHARM

– an eternal religion of nature in material world. It has no bias what so ever. It is based on basic laws of nature. This is based on VASUDHEV KUTUMBAKAM – Earth is one family. This does not pertain to humans only but to all living beings on earth. What a magnanimity of thought. There is no doubt that someday all world will follow this basic eternal, "Sanatan dharm". After such a downfall due to invasion of many invaders from all over the world, repeatedly. Civilization suffered culturally, economically and education and science-knowledge wise, but **the science of spirituality has sustained all effects of destiny. That was exactly un-understandable to invaders because of greed and sense gratification.** The Sanskrit speaking civilization has always nurtured nature. The VED and PURAN only studied and researched for absolute truth, and evolution of universe, spirit, soul etc. Based on this, decision to shun all materially based attitudes in the society. Defence system acquired negligible status. That became the main root cause of slavery, apart from other causes. Due to slavery every part of the philosophy and eternal DHARM was considered and taught as useless, ritualistic and orthodox. The mind set of generations of the country was made to feel that SANSKRIT SPEAKING CIVILIZATION is illiterate, religious, ritualistic and orthodox. Believer in worship of idols and deity. Forcing conversions and enslavement and torture of population, practiced by the invaders. In fact, the universities like NALANDA, TAKSHSHILA, KANCHI, VIKRAMSHILA etc. destroyed to gain literary and scientific supremacy. Not only this but the brutality against the DHAKA MALMAL (finest weaving technology of finest cloth) weavers to gain supremacy of business of products of invaders. NALANDA University was the world's first residential university having 10,000 students in the university. It had nine story libraries. During attack by Muslim invader, burned the library having 9 Million manuscripts which took 3 months for fire to extinguish (vide Wikipedia). The Sanskrit language was so scientifically developed and flawless. No other language has achieved that level till date. Alphabetic grouping is based on the science of phonation and the technical way the words are spoken. Reader is advised to refer details elsewhere.

The question answer form of GITA between ARJUN and Lord KRISHNA has cited examples which were quoted to emphasize the points. It will be discussed in all the chapters of this book. Description in GITA is highly scientific. Western world has started in believing in behaviour sciences. Hence GITA can be considered as PSYCHOANALYTICAL treatise also apart from scientific. This is how the mindset of persons and the society at large can be directed and improved to ultimate human benefits. Enmity, opposition, pride, deceit, anger etc. kept in mind but outwardly soft spoken, merciful and artificial honesty etc. if pretended by a person, is always against transcendental devotion to god – super soul. Person will not be benefited by such false behaviour. These things have a long-drawn effect and carried forward from generation to generations and incarnation to re-incarnations. The theory of reincarnations is not believed by materialistically involved in material world. Good deeds and bad deeds of living bodies and if soul is free from the bodies how then the deeds performed by bodily activities get with reincarnation? This phenomenon is unexplainable. The intriguing dilemma has been answered in above paragraphs in Gita itself. If one gets involved with the results of the work done, then attachment to material world is strong, otherwise one is free from bondage. Similarly, at the time of death the bondage of body to materialism persists and carried forward and next incarnation is provided according to the past deeds explained above. Because of this we see some persons enjoy happiness and get all good facilities of material world while other is full of miseries. Luck plays the role in that situation. What is this luck factor? This depends upon the deeds of past incarnations. It is a well-controlled and regulated law of destiny. Interdependent on each other deeds, KARMA and destiny.

In other words, LAW OF KARMA is applicable in destiny. **Karma and destiny work together. In fact, VEDIK scripts believe that one's destiny is unchangeable. It is linked with your KARMA. This is a divine law of cause and effect.** The fruits of our actions depend upon our deeds. In contemporary life we see that even if a person performs righteous actions and deeds, but he suffers bad effect. Why? Because our elements of past deeds play a part and suffering becomes unavoidable. On the contrary some persons act treacherously but spared from bad effect. Consequences of past deeds, (or KARMA), are experienced by all of us. Nobody is spared from this effect. In fact, there is no law that our good deeds will wipe out our past KARMA effects or KARMAPHALA. This is for the future effects. How it works is intriguing. Possible explanation given in VED and PURAN is, stored memory in subconscious mind (CHITTA). KARMA results show in future life or time. Bad KARMA performed in current birth/life by a person may not suffer because the good KARMA of past life may be showing fruits, but future life will be affected for sure. In later chapters of GITA, LIFE CYCLE of universe has been explained. Scientific explanation will be discussed appropriately in later chapters.

BRAHMA, VISHNU and SHIV are in fact not the same as painted in modern pictures or photographs, scientifically, these are part of cosmos, and a part of planetary systems like galaxies and stars etc. It has been described in space science, Astrology and Astronomy. SHIV being the main source and supposed to have been CREATED BY ITS OWN, self-existing or self-manifest. According to SHIVPURAN, RUDRA shiv is in the form of LING which has fire, heat and light energy equivalent to thousands and thousands of SUNS in combination. Many nuclear blasts occur frequently as occur in SUN of Solar system. These blasts are responsible for throwing massive round GAS BALLS which in due course of time become galaxies and planets etc. (explained in BRAHMOTPATTI in later chapters). Union of two elements e.g. Hydrogen and Oxygen gas with help of heat (burning) forms new element water. Union with heat water converted into steam and union with cold forms water drops again. As per Vedik literature, the Living beings are the combination of RAYI (Cosmic rays from various manifest cosmos – stars, Nakshtra, galactic establishments) and Pran (pranik energy). Details are discussed in later chapters of this book. Atom is union of particles (vide Partical physics) electron proton etc. But separation or fission generates tremendous energy (Atom bomb). The whole universe is union based.

Hence the ultimate source of universe is the outcome from a "collective energy source" – TEJAHPUNJ How Sages, Rishis have come to this conclusion is unknown. But sounds correct from the accurate mapping and predictions of the celestial drama in space e.g. eclipses, comets, stars and their names, Nakshtra etc. Stars can be accurately understood and seen from "JANTAR MANTAR" (for gazing celestial effects) constructed and still present in some cities of India even today. Probably, it is assumed that, the transcendental meditation and spiritual development may have played a great role in acquiring this knowledge. Suspicion comes to mind that whether they had some tools or gadgets to come to this conclusion. The SANSKRIT speaking civilization was so advanced that they knew all this scientifically. The modern civilizations too, are advancing and creeping to the same goal slowly. GITA is in the form of questions and answers, explained with examples. Generally, examples are cited which are understood by the questioner and depend upon the knowledge prevalent in that society. This means and can be concluded that the society during MAHABHARART ERA was certainly knowledgeable and the scientific knowledge was much more advanced than today. (The description of super soul, (or SHIV), is appropriately narrated in the MARATHI poem composed by author quoted in later chapter 11 page 100).

Keeping in mind after reading GITA by Lord Krishna to Arjun, Oppenheimer, father of Atom bomb, an atomic scientist, solaced his mind that work is worship, KARMA YOG is important. As a scientist his duty is to develop a bomb. To use THE BOMB or not, is work of politicians. His work as scientist is to develop science. This is without any attachment to its results. KARM IS MY DUTY, my right and not outcome. Leave it to GOD. It will not be out of context to elaborate here the MAHAMRITYUNJAY MANTRA from YAJUR VEDA, also called RUDRA mantra. This speaks about a melon growing on vine and ripening spreads sweet fragrant smell and on full maturity separates from vine and attains liberation – MOKSHA. As it has been described above. This is a natural process. It goes on with automation of nature. In case of humans, nature has provided humans with brain, mind, CHITTA – subconscious mind. To perform duty or otherwise depends upon his thought process. Performing duty is a decision taken by thoughtful humans only. Therefore, Lord Krishna has advised to Arjun in the form of GITA. This is for all humans and not for Arjun only. Such is the magnanimity of GITA.

MIMANSA (is basically curiosity) in VEDANT DARSHAN needs some explanation here as it deals with Work action (KARM KAAND) is POORV MIMANSA and Knowledge (DNYAN KAAND) is OOTTAR MIMANSA. Curiosity about and in basic nature of substance/elements i.e. Dharm of substance. Work or action done as per vedik suggestions to unfold natures of everything is included in YADNYA and MAHA YADNYA. Maha Rishi MUNI speaks the same:

Mahayadnyeshcha yadnyeshchbrahmiyam kriyate manuh

Work actions as YADNYA and Maha Yadnya only develops intellectuals like BRAHMINS. These KARM are of 5 types: 1. BRAHM YADNYA, 2. DEV YADNYA, 3. PITRU YADNYA 4. BALIVAISHWADEV YADNYA 5. ATITHI YADNYA.

a. **BRAHM YADNYA:** Self study – during joint period of NIGHT and DAY and Day and Night.

b. **DEV YADNYA:** HAVAN performed during morning (early morning) and evening. HAVAN is not a religious ritual (common meaning as understood in english). It is to enhance energy of human body to make it healthy and progressive. It is for mental peace and well being. It has therapeutic value in neurological ailments like Epilepsy.
    (Journal of EpilepsyResearch, 2015, 5(2: 33-45). (http://j-epilepsy.org). by P Bansal, R. Kaur, V Gupta, S KumarR. P Kaur. Doi:10.14581/jer.15009).

    For interested reader ingredients for Havan Samagri given below
    Therapeutic mechanism of action and active constituents of different components of Hawan Samagri on epilepsy

S.No	Name/botanical name	Active component Mechanism of action
1.	**Saffron Crocus sativus Crocetin**	picrocrocin, safranal, isophorone, 2, 2, 6-trimethyl-1, 4-cyclohexanedione, 4-ketoisophorone, 2-hydroxy-4, 4, 6-trimethyl-2, 5-cyclohexadien-1-one as well as 2, 6, 6-trimethyl-1, 4-cyclohexadiene-1-carboxaldehyde Increase in seizure threshold. Block PTZ induced convulsions. Increase GABA-ergic neuro transmission. Inhibit absence seizures. 46, 47 Improve tonic clonic seizures.
2.	**Jatamansi Nardostachys jatamansi Valeranone**	Calerene, patchouol, α-gurjunene, aristolone, β-maaliene, spathulenol Increase in seizure threshold, Inhibit the electroshock convulsions. 55 Increase GABA, 5-HT, 5-HIAA. 57
3.	**Coconut Cocos nucifera**	Monounsaturated fatty acids, Saponins Inhibit PTZ induced convulsions. Increase GABA level, serotonin level. 76 Anticonvulsant. 78
4.	**Sesame seeds**	Sesamum indicum 1-(5-methyl-2-furanyl)-1-propanone, 3-formylthiophene, 2-propyl-4-methylthiazole, 2-ethyl-4-methyl-1H-pyrrole, 2-ethyl-6-methylpyrazine, 2-ethyl-5-methylpyrazine, 4, 5-dimethylisothiazole, 4, 5-dimethylthiazole, 2, 6-diethylpyrazine, 2-ethyl-2, 5-dimethylpyrazine, 1-(2-pyridinyl) ethanone, and 1-(1-methyl-1H-pyrrol-2-yl) ethanone Decrease ROS, MDA in epileptics. 80
5.	**Clove Eugenia caryophyllus**	Eugenol, acetyl eugenol, β-caryophyllene, vanillin, crategolic acid, tannins, gallotannic acid, methyl salycylate, flavonoids eugenin, kaempferol, rhamnetil, eugenitin and triterpenoids like oleanolic acid. Increase onset of convulsions. Reduce duration of convulsions. Delay onset on seizures. Increase GABA ergicandglycinergic activity.
6.	**Nutmeg**	Myristicafragrans Myristicinandmacelignan Inhibit seizures. Reduce severity of seizures. 58
7.	**Nagkesar Mesua ferra Sesquiterpene**	diterpenes, triterpenes, carboxylic acids and saturated hydrocarbonsReduceHLTE. 61 Inhibit MES induced convulsions. Increase the onset time of seizures. Decrease duration of seizure.
8.	**Tagar Valeriana wallichi Valerian**	valipotriates and GABA sesquiterpene, diterpenes, triterpenes, carboxylic acids and saturated hydrocarbons Sedative action. Decrease HLTE. Anticonvulsant activity. 64

Contd...

S.No	Name/botanical name	Active component Mechanism of action
9.	Agar Aquilana malaccensis	Sesquiterpenes, benzylacetone, guaiene, anisylacetone and chromone derivatives Sedative action. 62
10.	Nagarmotha Cyperus rotundus Cyperone	selinene, cyperene, cyperotundone, patchulenone, sugeonol, kobusone and isokobusone, pinene (monoterpene) derivatives of sesquiterpenes such as cyperol, isocyperol and cyperone. Anticonvulsant action. 70, 71
11.	Ber Zizphus jujuba Flavonoids	saponins, tannins, vitamin A, vitamin B, sugars, mucilage, calcium, phosphate & iron. The pulp contains moisture, protein, fat, carbohydrate, calcium, phosphorus, iron, carotene, thiamine, riboflavin, vitamin C. Anticonvulsant action.
12.	Phoolmakhane Nelumbo	nucifera N-nornuciferine, O-nornuciferine, nuciferine, and roemerine, protein, amino acids, unsaturated fatty acids, minerals, starch, and tannins. Decrease tonic extensor convulsions. Anticonvulsant action.
13.	Mango Mangifera indica	PGG, polyphenolics, flavonoids, triterpenoids, mangiferin, catechin, isomangiferin, mangiferin, alanine, glycine, γ-aminobutyric acid, kinic acid, shikimic acid and the tetracyclic triterpenoids cycloart-24-en-3β, 26diol, 3-ketodammar-24 (E)-en-20S, 26-diol, C-24 epimers of cycloart-25 en 3β, 24, 27-triol and cycloartan-3β, 24, 27-triol.

c. **PITRU YADNYA:** Worship, gratitude and Devotion to Parents and revered teachers.

d. **BALIVAISHWADEV YADNYA:** To keep separate food out of cooked food for use of other living beings like other animals.

e. **ATITHI YADNYA:** Respect and welcome of guest coming home.

Gita says about YADNYA:

Yadnyarthat karmanonyatra lokoyam karmabandhanah I
tadarth karma kounteya muktasangah samachara II 9 II

(3, 9 GITA).

Actions/work performed as sacrifice (yadnyarthat reiterates the meaning of YADNYA refered above) for GOD liberates (frees) performer from the results or outcome of actions, and bondages of material world. (God is Lord VISHNU – Shri Krishna is in fact incarnation of Lord Vishnu, therefore sacrifice to GOD is to Vishnu or Shri Krishna or Super soul) If not it causes bondage in the material world. Hence perform your prescribed duty for the satisfaction of God. In this way you will be free from bondage. (Oppenheimer's example is appropriate here as he solaced himself on the preaching of GITA, as stated above). Work done, good or evil by any other way in this material world results in bondage. The work reaction binds the performer. Performing such action and activities with devotion to God, the performer is in a liberated stage. No work need be performed for sense gratification. Everything be performed as a duty for the satisfaction of God. The reaction of work will be bond free. This will slowly elevate performer to transcendental love and affection and be directed towards liberation.

Sahayadnyah praja srishtvapurovacha Prajapati I
anen prasavishyadhvamesha vostvishtakamadhuk II 10 II

(Gita 3, 10).

In stanza 3, 10, it is understood that, togetherness (सहयज्ञ:) of population and creator mutually for welfare of generations of populations granting prosperity and desirable material things by the bestowed with who bestows. **Creation of nature is by interaction of properties of elements in the universe. In due course of "time factor", newer and newer materials are created in material world. (see stanza below).** *This is a blessing of creator for prosperity and progress of generations of population. Because of all these descriptions the VED are called PRERPETUAL and eternal (सनातन) the subject dealt is perpetual. It is neither a cult, faith nor religion as in modern time. Universal existence, its creation and creator is for happy living and ultimate liberation of soul/energy and union with super soul. Liberation and unification with CREATOR (योग). Nothing is in hands of the living creature. Living beings life, birth or death, is not governed by the creature himself. It is controlled by the*

Nature (prakriti) and in turn by creator who has created all this nature and governed by automation. Hence HIS existence can be experienced, felt and assumed. HE is a charioteer (details in **Chapter 15 – brahma bhavati sarathihi***).*

The part above in italic is endorsed by the stanza of Gita 3, 28:

Tatvavittu mahabhavo gunkarma vibhagyoh I
gunaguneshu vartanta iti matva na sjjate II 28 II

(Gita 3, 28).

Knower of truth knows that acts of essence of elements, separately react. Modern scientific example for reaction (get engaged) of essential qualities of elements, can be quoted here is a chemical reaction of Oxygen and Hydrogen in a particular quantity along with heat and "time factor" (Kaal – KANAAD RISHI), forms water. The essential quality of Oxygen reacts with essential quality of Hydrogen. These react with each other in a time frame and form water. $(H_2 + O_2 = H_2O)$.

Regarding time factor the best example is coal becomes diamond after long lapse of time. Certain time is needed for the reaction to take place even formation of water cited above is not free from this time factor.

A scientific description on living bodies, Rain, Food, sun light, moon light, water etc. and their dependence on each other has been perfectly narrated in next stanza. Living bodies depend upon food. Food production depends upon rain, light, water, nitrogen cycle etc. in fact PARJANYA is thunderous clouds, whose production depends upon other factors (par=other Janya=birth, production). Production depends upon other factors. It is a perfect science as per modern science is concerned. It speaks of great science hidden in depth in the stanzas of Gita. Rain Indra devta) from clouds, clouds from sea and sun, and nitrogen cycle (pushna chakra,/ (one of the 12 names of sun god is PUSHNA), which is necessary for food production. Most important is YADNYA KARM, i.e. each and every element concerned with such a nature bound act, performs prescribed duty as sacrifice (yadnya karma), then only such a cycle is possible (samudbhava). See below:

Annadbhavti bhutani parjanyat annasambhavah I
yadnyat bhavati parjanyo yadnyah karmasamudbhavah II 14 II
(GITA 3, 14).

Still interesting and similar description exist in, in BHRIGUVALLI section, ANUWAK 2, of TAITTARIYE UPNISHAD. Rishi BHRIGU went into penance and found out that: food is BRAHM. Living beings are created from food. Live on food. And become food ultimately. What an observation e.g. Living beings are born due to food. Grows and live on food. And is a food for carnivores or bacteria ultimately.

Sa tapotapyat sa tapastaptva anna brahmeti vyajanat I
annat ev khalvimani bhutani jayante annen jatani jeevanti I
annam prayantyamisamvishntiti

In next chapters a description of Nitrogen cycle has been described, it will be discussed in appropriate chapter. Summarily it is described here:
    A verse from RIGVED clearly talks about Nitrogen cycle (pushna chakra).

Pushnachakram na rishyati na koshovpdyate I
no asya vyathatepavihi II 3 II
(Rigved. Mandal 6 sukata 543).

This clearly indicates that Nitrogen is maintained in the atmosphere. Neither it is reduced nor the store, is reduced. At the same time, its ability to work does not gets affected. This is essential for vegetation on earth. This is essential for the animal kingdom. For production of seeds moon rays are essential. All this is as per the modern science. Gita speaks the same in chapter 15, stanza 13:

Gamavishya cha bhutani dharmyahamojasa I
pushnami choushadhihi sarvah somo bhutva rasadhtmakah II 13 II
(Gita 15, 13).

Sun, Moon, Water, Air, light etc. are described as god (DEVTA). God or DEVTA means natural/supernatural elements. This should not be misunderstood as mythological or religious but be understood scientifically. Only then scientific facts can be understood. Unfortunately, society with modern mind-set has taken it as religious, when it has nothing to do with religion, faith, cult, SAMPRADAY. It deals only with nature and supernatural reality. That is what the modern science deals. Another factor which needs emphatic mention is OM NAAD. Big bang theory of modern science accepts creation by boom (OM) sound. It will be discussed in appropriate chapter. Thunder of cloud probably forces dissolution or attachment of **life forces**, present in ether in space, in rainwater drops which are present in ether in space around globe of earth. With rain it falls on earth and mixes in soil which nourishes life in plants (GITA 3, 14). Plants are primary producers. The animal and humans get all nourishment from plant food. VARUN DEVTA after dissolution in (rain) water, satisfy PUSHNA DEVTA. It will be wise to recollect that one of 12 names of Sun, is "PUSHNA DEVTA". Here it is important that do not misunderstand DEVTA as god and avoid religious tinge to it. The confusion is due to misunderstanding and diluting the whole spirit of science in SHRIMAD BHAGWAT GITA. In fact, this is the YADNYA KARM performed by all elements, DEVTA. This is responsible for rain, light, water, life forces etc. to produce FOOD. In this context interestingly RISHI PARASHAR developed meteorological science as VAARTA SHASTRA. It is regarding Agriculture, Meteorology and Commerce. This Shastra explains life under soil, Bacteria and insects etc. This is violence on life in the soil while ploughing. But this act is not evil because food produced offered to god, guest, visitor first and then to be consumed (Muchyatetithipujanat). To produce food for life to live is a pious work and is as per DHARM (not religious) of house holder (GRAHASTH). Bad effect of such work gets obviated:

Krushirdhanya krushirmedhya jantunamjeevanam krushihi I
hinsadidoshyuktopi muchyate tithipujanat II 8 II
(RISHI ParasherI. 8).

Similar opinion is expressed in MANUSMRITI 3. 71.

Panchetanyo mahayadnyanna hapayati shaktitah I
sa grihepi vasannityam sunadosherna lipyate II 7 II.

Householder is free from five places of violence when used like Fire stove (CHULLAH), Grinder (CHAKKI-), Broom, Mortar Pestle, POT for storing water He has to perform five YAGYA. Details are beyond the preview of this book.

How minute thought is given to all details of science. It is clear, that life exists in soil and under the earth in macro form and in micro form also.

It is emphatically stressed here that the examples cited were known to all people which explains prevalence of scientific knowledge in common public of that time, otherwise examples would not have been quoted.

It is said that the VED are directly manifest from supreme soul consequently the "all – pervading life energy the "soul". It is situated in acts of sacrifice. *NIRGUN NIRAKAR BRAHM cannot write VED but under transcendental meditation, whatever has come to minds of RISHI MUNI and sages of past, has been scripted down in the literature form.* Hence it is labelled as APURUSHEY – (not by man). There is a hidden science element in this stanza.

Karma brahmomodbhavam viddhi brahmaksharasamudbhavam I
tasmatsavarvgatam brahma nityam yadnye pratishthitam II 15 II
(Gita 3, 15).

Elements have physical form and chemical composition. The acts of elements exist due to soul and soul exist due to eternal (akshara-unending) supreme soul (parabrahma). Therefore, it is all pervading (sarvagatam\) Hence it always exists (pratishthitam\) in actions (yadnya). Therefore the production of food in nature is as per the laws of nature and is governed by constitution of super soul with the help of elements and automation as stated above.

The necessity of work for the satisfaction of YADNYA PURUSH which is DEVTA/KRISHNA/VISHNU is YADNYARTH KARMA (meaningful work) is stated in chapter 3, 15ᵗʰ stanza is in fact a working code or directions to work. Work done like this is not sinful and one is not booked for the reaction/results of work. In "BRIHADARANYAKOPNISHAD" it is stated that supreme soul is OMNIPRESENT (tasmat sarvagatam) and material nature received fraction of soul in every particle of material world. Supreme soul directed the eternal YADNYA KARM. This is the facility provided that after satisfying sense desires the fractioned souls return to supreme soul by following VEDIK KARM and attain liberation (MOKSH). Lord KRISHNA gives his own example, being a super soul Krishna is not bound by any obligation to perform duty nor he desire or want anything but he is engaged in prescribed duty. He is not bound by anything in all the three abodes of universe (three planetary systems = BRAHM lok, VISHNU LOK, SHIV LOK. But still he, as a common man, performed all duties prescribed by nature (DHARM) as one of the person in the material world (Gita 3, 22).

Na me parthasti kartavyam trishu lokeshukinchana I
nanvaotamvaptavyayam vrta ev cha karmani II 22 II

(Gita 3, 22).

This duty in fact, is to set an example for common man to follow. (GITA 3, 21). Because the way the leaders, elite or celebrities behave the rest of the people follow them.

Yad yadacharati shreshthastattadevetaro janah I
sa yatapramanam kurute lokastanuvartate II 21 II

(Gita 3, 21).

In this stanza, 3, 12 cites example knowing that it was the common knowledge in that society of SANSKRIT speaking civilization. Details are not given in GITA. Space science was highly developed. The details are available in MADBHAGWAT and other PURANS. In Chapter 8 details of life and duration of life span of BRAHMA LOK, VISHNU LOK and SHIV LOK has been described as example only. Details are elsewhere. This will be discussed in appropriate chapter. 7th Chapter of GITA stanza 4 & 5 the nature is divided into two groups PARA supernatural –soul, and APARA PRAKRUTI– material world. These are:

EARTH, WATER, FIRE, AIR, SPACE, MIND, INTELLECT, EGO, this is 8 types of APARA which is material world and the other is PARA which is super natural –soul. Whole world is under the influence of these two. The elements (TATTVA) or DEVTA, and the meaningful work (YADNYA) undertaken yields food production. This provides food grains for livings beings (already explained above). Same is in the stanza GITA, 3, 14.

(In the beginning of 20th century W. Ostwald, a scientist, explained that the matter in fact is complex form of energy. Aristotle – 384-322. BCE, matter is made of earth, fire, air and water. Democritus – 460-370 BCE Atoms are the essence of all matter.)

(Science Magazine – OYLA 13 January 2020. A Brief History of Matter. Page 4 to 13).

According to MAHARISHI KANAAD, (7000 B.C.) THE 8 MATERIAL TYPES OF NATURE, are all considered matter and of the entire universe. Earth is solid matter, water is liquid matter, Air in gaseous form of matter. He considered even the brilliance as matter. He goes further to say space and time are also matter. Modern science now started considering matter and energy as one (see above Chapter 2 page 38-page 89).

Pruthvyapas tejo vayur aakasham kalodigatma mana iti dravyani

PRITHVI, AAP, TEJ, VAYU, AAKASH, KAAL, DISHA, AATMA, MAN, ITI DRAVYANI

He expressed that Space, sky or AAKASH, is also matter which is without atom. Because o particles are present in space, therefore movements can take place. Hence aakash is **influence zone** between atoms.

Special imagination about MATTER by Maharishi Kanaad the "Subject" is named as VAISHESHIK DARSHAN. In this philosophy (DARSHAN) it has about 370 Stanzas or SOOTRA in 10 Chapters. The meaning of VAISHESHIK is which indicates "SECRETS of matter". Common and special virtues of matter are discussed. He says:

Dharma vishesh prasudad dravya gunakarma samanya vishesh samavayanam padarthanam sadharmyaveidhrmyabhyam tatvadnyanannihishreyasam Vaisheshik Darshan 1. 1. 4. And 1. 1. 5.

All matters are composed of small particles (Partcle Physics of modern science). Qualities and action of matter are its virtues (DHARM).

1. Earth can be recognized by Smell. Other virtues are Roop (Shape) Juice, Smell, touch (Sparsha). Main is Smell.
2. Water, can be recognized by coolness. Other virtues are Juice, Roop, Touch. Main is Juice.
3. Fire can be recognized by heat. Other virtues are Roop and Touch. Main is Shape.
4. Air can be recognized by special touch. Atoms are present in it.

Living beings body is made from these cosmic matters.

5. Aakash or Space can be recognized by Sound (SHABDA).
    All 5 are collectively called as five elements – PANCHBHOOT. The qualities of these are Smell, Juice, Shape, Touch, and Sound. The senses are Olfactory, Taste, Vision, Skin, and Listening.

6. Time-Kaal is imaginary. Past present future, Hour, Minute second are for practical understanding.
7. Direction (Disha) East, West, North, South, Above Below etc.
8. Soul (Aatma) can be recognized by Knowledge.
9. Mind. Pleasure, sorrow recognized by mind. It is a sense organ.

About Aakash the stanza says:

Black sky (ghatakash), Massive sky (mahakash), and sky of Heart (hridayakash).

(Vaisheshik Darshan – by Shri Narayan Mishra, Varanasi 1968).

In sky or space of Heart the spirit – AATMA dwells. Heart is anatomical organ but functionally it is a miracle. Muscles of heart is never fatigued till term of life it works 24x7 nonstop. Soul energy or "Aatm – Shakti" dwels in heart, brain, and whole body.

Here, in the same context, it is interesting to know the word HRIDAYA in Sanskrit. How the heart or the HRIDAY word came into existence. Word itself denotes the functional aspect of the organ HRIDAYA. In SHATPATH BRAHMAN (about 7000 years old) a description says:

hriter dadater yaterah hriday shabdah
(in split form of this stanza= Hriter, Dadater, yater, hridaya shabda)

(The functional origin of the word HEART or HRIDAY) explained below):

i.e. Hriter = receiving, Dadater = giving, yater =propelling, makes the word HRIDAY. Still more expressive stanza describing actions of heart:

Sankocha cha Vikasam cha svtah kuryat punah punah

Sankoch = contractions, Vikaas = expansion, Dilatation, Swatah = self, Punha punha = and done again and again.

All is self – explanatory. This explains that even the nomenclature and derivation of word is scientific. To emphasize here that the alphabets of the Sanskrit language is based on phonation. Kanthasth – originate from throat (or kantha), from lips othasth = lip origin = pa, fa, ba, bha, ma and so on.

(for details see chapter 19, page191-208, Indias glorious scientific traditions. Suresh Soni, 2010. Oceon Books Pvt Ltd.)

In next few stanzas Lord Krishna explains how the enemy in the form of LUST, influence the important places of the body of living entity of material world. Mind is the sole commander of sense. Mind is store house of all ideas when one listens about sense gratifications. Then lust influences the intelligence. With mind, intelligence and sense, the spirit soul is lured to enjoy the sense gratification. This is false and fictitious gratification and ego, mistaken for true happiness. Thus, the mind becomes enemy if left unsupervised or uncontrolled. The mind is super controller of actions and sense gratification. In other word the enemy slays the beholder of knowledge and self-realization if not taken care of.

Tasmatvamindriyanyadou niymya bharatshabha I
papmanam prajahinhyenam dnyanvidnyanam nashanam II 41 II
(GITA 3, 41).

After recognizing the enemy hideouts, regulating senses, mind and intelligence, one can slay the hidden enemy for getting true knowledge and self-realization. This is as per the maharshi **kanaad's statement stated above, that the energy, mind, and intelligence are all belong to matter category. The GITA in comparative form, states here that working Senses are superior to dull sense matter, mind is higher than sense, and intelligence is still higher than mind and the soul is higher than intelligence**. Therefore, soul is the solution of all the problems. With spiritual intelligence and strength conquer the enemy in the form of LUST. Controlling mind and insatiable desire – lust as enemy can be conquered.

Evam buddheh param buddhva sanstabhyatmanmatmana I
jahi shatru mahabaho kamarupam durasadam II 43 II

(GITA 3, 43).

One can realize that the science is hidden in stanzas of Gita and must be explored with reference to scientific explanations as per science facts today, for the understanding of current population. Science prevalent in the society during MAHABHARAT PERIOD, was used by Shri Krishn to convince Arjun. In this stanza it is clearly indicated that the sense enjoyment and gratification is enemy and needs to be slayed by spiritual intelligence as this enemy prevents conscience and scientific achievements of the material world. Physical material world is camouflaged by luring senses of the material world, which always demands gratification and enjoyment. To gain satiety trapped embodied soul can go to any extent. In reality it is false and grievous. What is superior in material body is explained in chapter 3 stanza 42 is senses are superior to body, mind is superior to senses, Intellect is superior to mind, while soul is superior to intellect. Hence with spiritual intellect and transcendental mind-set control, the science of material world can be achieved. No doubt with such a preaching to population of society all round progress was achieved by Sanskrit speaking civilization.

HISTORY OF DIVINE SPIRITUAL SCIENCE TRANSCENDENTAL KNOWLEDGE SCIENCE OF BASIC DHARM AND PRAN
(LIFE ENERGY)

Abstract: History of Gita is explained. Even after expense of a huge span of time Lord Krishn tells Arjun that the divine spiritual knowledge is being retold. This is due to number of birth and death of Arjun and Shri Krishn. Is there a science of birth? The original stage of Nature is capable of giving birth, to a new life. The subtler form is essential for body formation. (explained in previous chapter). Using internal energy, the supreme soul with help of nature can take birth, even though 'HE' is unborn and not deteriorate-able. This way in "I incarnate my self". This is responsible for the divine spiritual science. This unfolds possible history of Gita and existence of transcendental material body in manifest world, repeatedly present. It is since time immemorable. There is indication of GOD particle on page 118. It needs to be researched out.

Lord Krishna himself narrates history of GITA in the beginning of this chapter. Science of relationship with supreme soul is not new. The eternal YOG knowledge, culture and education were imparted to god of suryadev "vivaswan" who in turn to manu (father of mankind).

It is statement in Gita that all Humans are from a common source. Interestingly modern science is inching towards the same conclusion. Genetic researchers are engaged in finding the source of all human race. The fact is still questionable. Various researchers all over the world seems confused on their scientific success. But How Gita, Vedik literature, could tell emphatically the MANU is father of all humans?

The conclusion, of modern investigators, is awaited. (see below):

A new **genetic** study suggests **all** modern **humans** trace our **ancestry** to a single spot in southern **Africa** 200,000 years ago.... "It would be astonishing if **all** our **genetic ancestry** at this time arose in one small homeland." Modern **humans** arose in **Africa** at least 250,000 to 300,000 years ago, fossils and DNA reveal. Oct28, 2019. Experts question study claiming to pinpoint birthplace? Refer web site: www. sciencemag.org>news>2019/10>experts-quest.

MANU later imparted this imperishable science to ISHWAKU (GITA 4, 1).

Eimam vivasvte yogam proktavanahamvyayam I
vivasvanmanave prahamanurikshvakve braveet II 1 II

(Gita 4, 1).

Surprisingly, it is narrated in TRETA YUG. Therefore, Science of GITA cannot be considered insignificant, speculative treatise but a knowledge eternal and from time immemorial. From archaeological, point of view, on earth, the Sanskrit speaking civilization has calculated time scale and periods (KAAL) into various groups. The number of years grouped into various Yug, named Satya Yug, Treta Yug, Dwapar Yug and Kali Yug. Number of years in these YUGAS are different. The Shantiparv of Shri Madbhagwat, history of GITA can be retrieved *(SHANTI PARV 348, 51-52)*. It states:

Tretayugadou cha tato vivasvanmanve dadou I
manushchalokbhrutyartham sutoyekshvakave dadou I
eikshvakuna cha kathito vyapya lokan avasthitah
(SHANTI PARV 348, 51-52).

The meaning is same as stated above. It is stated in SHANTIPARV as well. The science of KAAL (काल) or time is clearly stated elsewhere in chapter 8 of this book. The calculations of YUG can be estimated from Shrimad Bhagwat Puran, Brahma Puran and Markandey Puran. To calculate Tretayug, at present Kali Yug time spent is 5000 years, which has 432000 years. Before Kali Yug was Dwapar Yug which comprises of 800,000 years Before this it is Tretayug (1200,000 years), that means Manu spoke to Ishwaku about 20.05000 years before krisna told to arjun before starting Mahabharat war. Before Manu, the Gita was narrated to Vaivasvan. Hence Gita is considered as superhuman. The Gita was narrated to Arjun 5000 years ago at the start of Kaliyug when LORD Krishna Blew the mighty Konch Shell "Panchjanya" at the time of Mahabharat war in Kurukshetra. This book is not for the history of Gita but exploration of Science in Gita. From archaeological science it is a written history, about Gita and Mahabharat period. **From these calculations, one thing is clear that invention of zero seems eternal and not new. On zero depends the science of space and other sciences, which were well advanced before MAHABHARAT war on the battlefield of KURUKSHETRA.** It will not be out of context to mention here that the speed of light was known to Vedic civilization.

Speed of Light is calculated in Vedas (RigVeda 1.50.4)

"Tatha ca smaryate yojananam. sahasre dve dve sate dve ca yojane
ekena nimishardhena kramaman"

The meaning of "dve dve shate dve" is "2202" and "ekennimishardhena" means "half Nimish"/

This means while praising SUN it is said that the light coming out of SUN travels 2202 Yojan in half NIMESH.

"[O Sun] you who traverse 2,202 yojanas in half a nimesha". Distance Calculation:

1 yojana =9 mile, 1 mile = 1.60934 km

In TAITTARIYA UPNISHAD the stanza is:

Yojananam sahastre dve dve shate dve cha yojana I
eken nimishardhen kramabhananmostute II iti II

I salute SUN Rays whose speed is 2202 YOJAN in half NIMESH (Sanskrit scale is given elsewhere). It is 1,85,000 Miles in one second. As per modern calculations the speed per second is 2,99,792 Km. Conversion of speed in kilometer is 2,97,728 kilometer per second. One can understand the importance of zero in such calculations.

Another fact can be concluded is the teacher-pupil tradition this knowledge was told by one to another. There may not be any written statement available about GITA. The scientific knowledge might have been lost or chain of spoken knowledge might have been broken and lost.

Evam paramparapraptamimam raharshayo viduh I
sa kalneha mahato nashtah paramtapah II 2 II

(Gita 4. 2)

This science was told again by Krishna himself. It can be considered true because lost city of Dwarka, abode of Lord Krishna, has been found out by archaeologists of India. (see General considerations). In 3rd stanza of 4th chapter LORD KRISHNA reiterates to ARJUN that the ancient science of relationship with supreme soul is "narrated to you by me because you are my friend and devotee and you have the capacity to understand the mystery of this science". It is clear here that, the knowledge must be extended to only the deserving pupil.

Regarding the time scale Time cycle (kaal chakra) it is very clearly stated that the time is dependent on movements of Sun and planets with orders from super soul (see below). (See chapter 8) The SUN rotates in its orbit as per the order from supreme soul. A statement from BRAHMA SAMHITA:

Yachchakshuresha Savita sakala grahanam raja
samastasurasheshatejah I
yasyadnyaya bhramati sambhutkalchakro
govindamadipurushamtamaham bhajami

The scientific facts are evident in the stanza.

1. SUN is king of all planets.
2. The SUN god, king of all planets under control and supervision of VAIVASWAN gives heat and light and controls life and life processes in living beings, at the same time.
3. under supreme soul's control, moves time and time cycle is third scientific fact. Scientifically speaking the **Sun star is not stationary but moves, revolves and controls time cycle (KAAL CHKRA) is clear in this stanza. This statement is even before the narration of GITA to Arjun.** Hence Lord KRISHNA selected VAIVASWAN as disciple to educate with the science of GITA. In 2nd stanza GITA:

Evam paramparapraptamim rajarshayo viduh I
sa kaleneh mahata yogo nashtah paramtapah II 2 II

As stated above, the science was lost due to time lag and this traditional knowledge running in royal families KING and their kith and kin as rulers. But as per 3rd stanza, the same old mysterious science is being narrated to ARJUN by LORD KRISHNA. LORD KRISHNA speaks the famous stanza. Not only this, but the following stanza gives another scientific fact about sun:

Yen suryastapati tej seddhah I
navedvinmanute tam bruhantam I
(Teittariya brahman, 3, 12).

(quoted from Patanjal Yog Pradeep Gita press Gorakhpur, Page 13, "Shad darshan Samanvaya").

Power, strength, brightness, or sharpness of Sun is not of its own but is due to something else. It is bestowed by the Supreme soul to sun. Sun rotates and revolves in space is due to energy provided to sun by super soul. Modern science considers light and heat is due to repeated nuclear blasts. It is true as it is directly evident that the heat and light is produced due to blast. But blast and the material responsible for blast, are provided to sun at the time of birth of Sun Star. It will perpetuate till lifetime of Sun. At present it is quite on this aspect about sun. It will be revealed to modern science in due course of time.

Rotation, revolution, brightness, sharpness or strength of sun and emission of rays or explosion in sun or repeated solar flares in the sun is provided to sun by super soul becomes DHARM of sun. In this context it is reiterated here that: ("yada yada hi dharmasya" In the beginning of this book)

A common meaning of DHARM is religion. What is the meaning of DHARM in this stanza? What Vedik civilization meant by word 'DHARM". Everyone feels about religion only. If you ponder on this stanza and meditate a little you may realize that about 5000 years of bygone history how many religions prevailed during that period? Muslim, Christianity, Jain, Buddha, Hindu Sikh etc. religions never existed during that period of MAHABHARAT era. Then what is the meaning of DHARM in Shrimad Bhagwat Gita. Gita is the perfect presentation of VEDIK knowledge. Some stanzas are exact repetition of script from VED. Dharm refers to that which is constantly existing with a material substance, object, or element. If one discovers the essential part of a living being, or a part which is a constant companion of that substance. This has eternal quality, which has its eternal DHARM. MAHARISHI KANAAD very clearly had stated approximately, 7000 years ago that:

Dharma vishesha prasudat dravya guna karma samanya vishesha samavayanam I

padarthanam sadharmya veidhamaryabhyam
tatvadnyananihashreyasam

DHARM is as per matter's (Padarth, Dravya etc.) general action, work, virtues. It has nothing to do with the modern religion. Therefore, meaning of DHARM be taken in its real intention without any distortion. Unfortunately, the modern scholars of Indian origin also discuss Dharm as religion. Even instead of talking about SANATAN DHARM these scholars have conveniently described "sanatan dharm" as "Hindu religion" under the influence of invaders and rulers. It needs rectification in basic concept of DHARM and properly understood the meaning of dharm. To elaborate further with example following description is important:

The fire has heat and light without which it is not fire. In absence of this there is no meaning of fire. This is the SANATAN DHARM of fire and is not sectarian or a faith as the modern religion means. Thus, maintain form and function of the object. Perform settled function by nature which is required by nature of the object or living being. (*Here it is important to understand that everything in this universe is living. All universe is governed by birth, duration of life, growth or change and death. Time is another factor that govern living life. All stars of universe are born and die one day*).

Consider the stanza meaning if this integral DHARM of substance is disturbed the object may disintegrate and finish. Sectarian process of religion is faith. This meaning is certainly not DHARM at this place. What is meaning of sambhavami yuge yuge-

All round progress is DHARM of every human as said by MAHARISHI KANAAD (a philosopher scientist-approximately 600 B.C.-Exponent of atomic theory – which is known as "VAISHESHIKA SUTRA"). Definition of DHARM by Maharishi Kanaad:

ˈYatobhyudayat shreyasa siddhihi Sa Dharmaˈ

Physical/Philosophical, Spiritual all round progress is so called "real RELIGION" or DHARM. This means that the word used in this stanza is not with the meaning of modern understanding as religion. "Dharm" is eternal virtue. It had a different meaning. It needs to be understood in a correct perspective. It can be explained scientifically again as:

As per modern science, the laws of physics are applicable in this situation. Every structure, object in this universe maintain its form. How it is achieved? A cohesive force exists which binds all components of object/living/non-living (in fact nothing is non-living. As already stated before, everything in universe takes birth survives, grows and dies). This cohesive force can be considered as the DHARM of that object. This cohesive force can be electromagnetic force or Valence or Gravitational force of earth/stars, energy or equilibrium force. This is DHARM of that object of material world. Whenever this DHARM is lost disintegration process starts in the substance. The object gets shattered. To avoid this, equilibrium or electromagnetic force, valence needs to be re-established to avoid disintegration. According to Faraday's law of electromagnetic induction and Lentz's Law when the energy, force is void (ADHARM) and the space in between the particles of structure (objects are composed of small sized particles – **particle physics**, suggested by Maharishi Kanaad) to maintain structure and re-establish structure and equilibrium energy force must be infused. This is responsible for preventing decay or disintegration of structure. The energy appears in between the particles. The disintegration is stopped. In Gita Chapter 4 stanzas 4-8, Lord KRISHNA emphasizing that "I am involved in this process since time immemorial – eternal (Sanatan kaal). This is DHARMA STHAPANA. The subject of "**particle physics**" has received attention and developed in 20th century. The substantial part of Macro-cosm is matter as it is the basis of everything around us. Thousands of years before Maharishi Kanaad has mentioned in his book. How the stability of a substance is maintained by interaction of smaller particles of which the substance is made up of. The space between the particles of the substance allows freedom of movement of particles. The use of electromagnetic and gravitational force in substance is responsible for stability of the substance is known now. (Science of today, OYLA-

13th January 2020, page 4-8). But this was well known to MAHARISHI KANAAD thousands of years before is a big surprise. His statement is:

Prithvyapas tejo vayur aakasham kalodigatma mana iti dravyani

PRITHVI, AAP, TEJ, VAYU, AAKASH, KAAL, DISHA, *AATMA*, MAN, ITI DRAVYANI.

As per this stanza ATMA is also included in particle entity (kalodigatma-Kaal, diga (disha - direction), Aatma). Has he indicated this as GOD PARTICLE? This is what the modern science is searching for GOD PARTICLE. This needs more detailed research and analysis.

After convincing ARJUN about many birth and death and reincarnations of ARJUN and KRISHNA and recitation of GITA in the past, Lord KRISHNA states the famous stanza stated above. To explain further:

System of DHARM (religion) of universe (not Hindu, Muslim, Sikh, Christian or any other modern Religion, Cult, SAMPRADAY), and laws about the maintaining such religion depends upon the "Supreme lord", as neither birth nor death of living being, is in their hands but have to depend upon super natural activities of universe, even though nature has established everything under the laws of AUTOMATION. Example of this is when a child is born male or female both are having similar growth, the development of both is not different, both are same, but at puberty with automatically generated HORMONES make the real difference between male and female adults. DHARM of nature which is re-established by nature or supreme lord by the way of HIS AUTOMATION, is arranged by the SUPREME LORD. The supreme lord here in GITA is Lord KRISHNA. Therefore the words "I come again and again" ("Sambhavami yuga yugae"). YUG is a time scale mentioned in VED and PURAN many times repeatedly. Therefore, the supreme soul appears again and again from YUG after YUG – time immemorial to establish DHARMA of the living structures. This is mentioned earlier that GITA has been narrated to controller of SUN

– VAIVASWAN. YUG period has been mentioned earlier in this book. This incarnation of SUPREME SOUL in material world is known as AVTAR. This can be anywhere in the world any place, city structure or living being. The AVTAR can be in various kind, any form, shape, size or can be scheduled universally. BRAHMA SAMHITA says 'ANANT RUPAM' (INUMERABLE Form or RUPA) i.e. uncountable forms.

According to 16th SUTRA of 4TH chapter entitled Sankhya Aur Yog Darshan, Dharm is classified into 10 groups: Union of un-manifest energy and manifest nature is responsible for creation of universe. There are 10 original DHARM. DHARM present in manifest nature are of working type and DHARM in unmanifest super soul are original DHARM. As per SANKHYA and YOG DARSHAN, names of "ten Dharm" are (mentioned below), analyzed scientifically, responsible for the stability of a substance. These are:

Dash moulikarthaha ||16||,

(Patanjal yog Pradeep 4th Chapter, sankhya aur yog darshan, dash mool dharm, 16th stanza. Page 96).

<u>**Astitva, sanyoga, viyog, sheshvrutitva, ekatva,**</u> arthavatva, pararthya anyata, akartrutva, bahutva. (The highlighted and underlined five DHARM Provide stability to substance, element, matter, materials of the material world. First four are for manifest and un-manifest both. SANYOG, VIYOG are un-manifest's natural, effective DHARM). **Stability of matter depends upon these ten principles:**

1. The existence (ASTITVA).
2. The Coexistance (SANYOG).
3. The Disconnection (VIYOG).
4. The Balance sheet (SHESHVRITITATVA).
5. The Unity (EKATVA).
6. The Economics (ARTHVATVA).
7. The Affordability (PARARTHYA).
8. The Other (ANYATA).

9.   The uncollision (AKARTRITVA). Uninteractive or Catalyst.
10. The Inplenty (BAHUTVA).

The meaning of DHARM is understood by particular people under given circumstances. This depends upon the consciousness of people concerned or so to say development, intelligence and understanding. It is not the same in all individuals, is a well-known fact. Circumstantially DHARM word is taken as religion. In Sanskrit it is not religion at all.

Social science: Gita clearly indicates that different classes of people in society is created by Super soul. It is based on scientific principles. According to nature and virtues of material nature of work associated with people of society, four group of division of people of the society, depending upon the development intelligence and expertise of incumbent of society, has been established in human society (VARN DHARM). Different types of humans are created by the super soul. Gita has emphasized this in chapter 4, stanza 13.

Chaturvarnyam maya srushtam gunakarmavibhagashah I
tasya kartarapi mam viddhya kartaramavyayam II 13 II

(Gita, 4, 13).

This social order is created by KRISHNA (the super soul) as narrated in stanza 13 of this chapter. **Being a creator of this system,** still HE is not a doer of this system. Intelligent men are BRAHMINS, Administrators are KSHATRIYA, People with business aptitude are VAISHYA, while labor group are SHUDRA.

In this context description in BRIHADARANYAK UPNISHAD 1. 4. 1-7, the manifestation of population of living entities is thought and scientifically narrated. In the beginning of universe BRAHM was alone (1. 4. 1). He overcame his fear of loneliness (1. 4. 2). BRAHM desired and divided self into two (1. 4. 3), as unicellular living entities have this capacity e.g. Amoeba. (Laws of nature of universe have uniform applicability has been mentioned earlier).

sa va naiva reme, tasmādekākī na ramate; sa

dvitīyamaicchat | sa haitāvānāsa yathā strīpumāṃsau

samparişvaktau; sa imamevātmānaṃ dvedhāpātayat, tataḥ

patiśca patnī cābhavatām; tasmātidamardhabṛgalamiva svaḥ iti

ha smāha yājñavalkyaḥ; tasmādayamākāśaḥ striyā pūryata eva;

tāṃ samabhavat, tato manuṣyā ajāyanta ||3||

(Brihadaranyak Upnishad 1, 4, 3)

Every food grain has two parts in it, while germinating become two cotyledons-(dvitiyameichchhat, dvedhāpātayat), storehouse for food for growth of plant, is well known to all botanists (highlighted part in stanza above). This was with idea of expansion (PRAPANCH) of living entities in universe (not only on earth but whole universe). Various animal, insect forms manifested (1. 4. 4). In this way the union of two is responsible for the manifest world. By churning of mouth and two hands Brahm manifested FIRE. The juicy form, which is vegetation, was manifested with heroism/semen. He created better than self, i.e. being mortal himself he created AMRIT – Nector. This is excellent manifestation (Atisrishti – 1. 4. 5).

(BRIHDARANYAK UPNISHAD Code 577, page-162-187. Gita Press Gorakhpur)

These four classes of human society, were there but not strict as there were many examples of SHUDRA coming to other three groups, depending upon the person's aptitude, ability, development, intelligence, and work capability. To take advantage of the system invaders of country exploited the situation to their advantage for ruling and conversions to their CULT, faith etc. this continued for about hundreds of years. Generations passed and started believing the situation in the manner invaders wanted. It is absolutely clear from the News Paper cutting of 2nd February, 1835. Scrutinization of this newspaper cutting, is essential to understand the sinister design planned to enslave the peoples of the society. (see newspaper cutting, on page 103, Chapter 3.)

Here it will not be out of context that human society all over the world is divided into these four groups because as per Gita it is established by super soul (Krishna) page 133. CHATURVARN of

VEDIK description as stated in GITA exists all over the world. It has universal existence. It is not only social but highly scientific because it is based on the intelligence, ability, skill and development of individuals. A question can be asked whether that the "white house" chief occupant in America will ever sweep outside white house or "10, Downing street" of London, United Kingdom, occupant will ever sweep "10 Downing Street" and vice a versa? Infact the four groups are everywhere in the world. (Stanza 13 of Chapter 4 of GITA stated above). It is not worth the condemnation in any perspective. Needs to be understood scientifically. Condemnation of the system is due to vested interest of invaders.

This system exists even today universally. Material nature of an individual human in human society universally, decides nature of work undertaken by that person depending upon skill, intelligence, ability etc.

What is the science behind birth? It is explained in stanza 6 of this 4th chapter. It says:

Ajoyapi sannavyayatma bhutanamishvaropi san I
prakritim svamdhishtaya sambhavamyatmamayaya II 6 II
(Gita 4, 6).

Though unborn (Ajopi), un destructible soul, with use of NATURE (prakriti), body of living being (Bhutanamishvaropi) are manifested by the soul factor in this universe (Sambhavamyatmamayaya). Material, physical body is needed by all living beings. To elaborate further, the science of birth, the material body frame is required which is through parents. Parents are not needed by Amoeba, to multiply. It divides itself in two parts. The two parts become two different amoebae. Here the nature is used by amoeba to manifest in universe. The body of multicellular organism is made up of numerous cells. Cells have DNA. Life originates in DNA particles. Therefore, the nature (Prakriti) is responsible for origin of life. The existence becomes possible. This is the science of birth, in this stanza GITA 4, 6.

Coming back to the four social arrangement i.e. "CHATURVARN" of stanza 4, 13 stated above, the society has incumbent with different development mental, intellectual and spiritual etc. The scientific fact behind this theory of "Chaturvarn" is detailed here:

In society, all humans are not having same intellectual development. This depends upon the material nature of that person. A spiritual knowledge and development is directly proportional to intellect of the individual. CHERO (W. J. Warner. 1866 – 1936. Astrologer who studied astrology and Palmistry from INDIA. He stayed in Caves of India to study palmistry, from TAD-PATRA and TAMRA_PATRA (copper leaf) with RISHIS in caves). He is author of Cheiromancy and Chironomy, Palmistry for all. He has described in his book on palmistry – CHEROMENSY and CHEROGNOMY, seven types of palms of hand starting from brutal or animal like to philosophic hands, a highly developed human being. This division is based on intellectual development of the beholder of palm. Sanskrit speaking civilization had number of highly illumined Rishis and Sages. They with their YOGIK power could visualize the colour of aura around living bodies. Modern science could photograph the individual AURA by KIRLIN photography. Colour changes with the character and quality of life adopted by the individual. Four groups could easily be organized, this system is called as VARN system:

1. High intellectual caliber, culture and development of the persons, emit white light around them (whitish VARN or color) as AURA, who could easily understand supreme soul/Brahma, are called as BRAHMINS.

2. Some persons having good culture, tendency to fight for injustice, intolerant to dishonor are with red aura around them. Generally, of Martian character. (Red VARN – color). People of such aura are fighter and called as KSHATRIYA.

3. Some people develop tendency for trade and commerce emit yellow light as aura. Such people help to grow the economic status of the society and country. They are called as VAISHYA.

4.   Some persons emit dark light as aura (Dark VARN). Aura is somewhat disturbed. Such persons are generally underdeveloped. They have serving tendency. Hence called as SHUDRA.

This system was not strict and not a "watertight" system. It was interchangeable, depending upon the capacity of the individual. This has been narrated in MANU SMRITI. The individual persons CHITTAVRITTI and desire to upliftment or down fall in varn system can change the VARN status. It will be appropriate here to mention about CHITTA.

   There is no equivalent word in English, but "mind-set" is acceptable meaning of the word. This as per PATANJAL YOG SHASTRA it undergoes various stages or AVASTHA. Control of behavioural attitude of mind/CHITTA. Hence in YOG it needs to be understood:

Yogaschittavritti nirodhah II 2 II

(2, patanjal yog sutra).

means YOG is curbing of waves and ripples of mindset.

Various stages are:
Psychophysiological Science, behavioral science about mind set. This the mind-set is most important in Yog philosophy. To explain the mind set in short it is expressed into nine different phases/stages.

1.   Awake stage
2.   Dream stage
3.   Sleep stage
4.   Deep sleep stage (pralayavastha),
5.   Initial Samadhi stage
6.   Concentration stage (ekagrata avastha),
7.   Differentiation stage (Vivekakhyati),
8.   Seeing actual things stage (Svarupavsthiti),
9.   Existing in self stage (Pratiprasava avastha).

A yogi undergoes through all these phases before achieving SAMADHI.

(Patanjal yog Pradeep 4th Chapter, SANKHYA AUR YOG DARSHAN,. Page 134).

But humans having different development and stages of mind are:

MOODH avastha –**Foolish stage** – sleep, greed, fear, idleness, doubtful thought – low level human. This is at high priority in such personalities.

KSHIPTA avastha – **Sane stage**, grief, worry, unsettled mind, involvement in material world-ordinary level of human attitude

VIKSHIPTA STATE – **greater sane stage** – pleasant ness, forgive ness, aware ness, bravery, mercy-higher level of human attitude and behavior.

EKAGRA AVASTHA – **focused state** of mind – steady ness – yogi

NIRUDHA AVASTHA – **transcendental focus stage** – meditation stage-higher level of yogi.

(Patanjal Yog Pradeep page147 & 148,).

Similarly, human beings are also of different level of mental development. All humans do not have same level of intelligence and differ in their mental stage of development. Accordingly, CHITTA AVASTHA also differs. Humans are not different from animals of animal kingdom on this earth. But apart from satisfaction of sense satisfaction like animals, spirituality is the highest goal for the humans. Sages of Sanskrit speaking civilization set such high goal for the whole humanity. Liberation is the ultimate and must be the aim, of living beings. This is SANATAN (eternal, perpetual, for all the time and for all) aim and DHARM for human race of this earth. This is un-bias opinion of Rishi, Sages and Muni of Ancient Bharatvarsh.

Brahmarpanam brahma havirbrahmagnou brahmana hutam I
brahmeiva ten gantavyam brahmakarma samadhina II 24 II

(Gita - Ch. 4, Shlok 24).

The ultimate transcendental knowledge is spelled in stanza number 24. The work is "brahma karma". The work, work done, consummation of work and fruits of work, all is brahma, the actions needed to accomplish all, ultimately dedicated to brahma. This can be simplified by the example of foodstuff. One knows that the food item is basically Brahm, created by Brahm. When one eats the eating person or individual is Brahm. The food brahm is delivered to eating person who is brahm. The "HAVAN" stuff food as "HAVI" is put into digestive fire which is brahm i.e. brahm delivered to brahm. Action performed is brahm karm. This is an understanding if developed by a person becomes liberated and ultimately reaches brahma or becomes brahma. This is spiritual science and ultimate thing to be perfected by all human race – is the 'ultimate-aim' of human incarnation. This is the difference between human life and animal life. Animals live for Sense enjoyment while humans have many more goals to accomplish. This is not only advocated but how to achieve the goal, methods are described. For the advancement in spiritual realization, certain methods are described in GITA stanza 29. Viz. PRANAYAM. To analyse stanzas from 27 to 31 of chapter 4, the science of PRAN was searched. PRAN is of utmost importance in PRANAYAM. Hence YOG SHASTRA was searched:

Apane juhyati pranam pranepanam tathapare I
pranapangati ruddhva pranayamparayanah I
apare niyataharah pranan praneshu juhyati II 29 II

(Gita 4, 29).

This is best described by Patanjali Rishi. Details are described in a book Patanjal Yog Pradeep. Lord Krishn explains to ARJUN, with passing reference as example. It has been quoted in Gita, because PRANAYAM SCIENCE of breath, was well known since time immemorial. The techniques of control of breath was well known to peoples of the society. Breathing air in and out through nostrils, is act of breathing only. PRANAYAM is more than that. PRAN is not inhaling oxygen and exhaling carbon di oxide only, as modern science understands it. PRAN has been described as life energy or CHETANA. ETHER in space contains **PRAN or life energy**. To explain it, the act or art of breathing is highly scientific. One knows that the breathing if stopped by strangling for 3 minutes or more the person or animal dies. By forced inhalation and expiration if more air is inhaled and stopping inhaled air for some time inside lungs, the absorption will certainly be enhanced and PRANIK energy in more quantity will be assimilated in blood and blood circulation, will definitely carry more oxygen and PRANIK energy. The tissues and organs will get more rejuvenation. These are the scientific explanations of PRAN+AYAM. Rishis and sages did not stop here but divided PRAN into 10 different types. Their nomenclature and flow channels and function of each PRAN, have also been described.

PRAN has been described as of 10 types. Names of these ten types are narrated by Sage PATANJALI. Different types of PRAN have different work to be performed in all living animal bodies. These have different flow channels for PRAN to flow:

FIVE essential PRAN, described and named as: PRAN, APAN, VYAN, SAMAN, UDAN.

The other five names of the PRAN are: NAG, KURM, KRUKAR. DEVDATT and DHANANJAY.

All of these have different functions to be performed in the body. Kindly refer to main work elsewhere. In short it is in the stanza given below:

Pranopanah samanashchya dano vyanastheiv cha I
nagah kurmah krukarako devdatto dhananjayah I
pranadya pancha vikhyata nagadyah pancha vayavah

(page 228, Patanjal yog Pradeep, Samadhi Pad), (Dhyan Bindu Upnishad 56-57). (Quoted from PANJAL YOG PRADIP, Gita press Gorakhpur, page 228).

Apart from this the channels through which the PRAN (life force or energy) flows and circulates in body to required regions – tissues and organs have also been narrated. Ten channels have been described. The names of these channels are in the following stanza:

Pradhanah pranavahinyo bhuyastatra dash smrutah I
Ida ch pingala cheiv sushunmna cha tritiyaka I
Gandhari hastajivha cha pusha cheiv yashaswini I
Alambusha kuhuratra shankhini Dashami smruta I
(Dhyanbindupnishad 52-53. Yogchudamanyupnishad 16-17)

Names of channels are: 1. EIDA, 2. PINGALA, 3. SUSHUMNA, 4. GANDHARI, 5. HASTAJIVHA, 6. PUSHA, 7. YASHASWINI, 8. ALAMBUSHA, 9. KUHURATRA, 10. SHANKHINI.

First three Flow channels in Modern science are right and left sympathetic nerves and third is in spinal column starting from Brain stem down wards. The other channels could not be confirmed. The modern science has not researched out these channels yet. It is a subject of research for the newer generation.

(Quoted from PATNJAL YOG PRADIP, Gita press Gorakhpur).

First five are important and essential life energies.

1. PRAN is resting in heart, in upper part of body and regulates sense of upper portion of body.
2. APAN Vayu rest in rectal ad anal region. It regulates lower body senses i.e. evacuation depends upon this VAYU. This energy as APAN VAYU controls evacuation of gas only without stool! Similarly ejaculates only semen and not urine during mating of male and female!
3. SAMAN VAYU rest in central part of body in UMBILICAL REGION and regulates regions of heart to rectum by communication, infusion, or movement (Sanchar). Consumed food, water etc. equally distributed to all parts of body.
4. VYAN VAYU regulates blood circulation in all parts of body.
5. UDAN VAYU controls body in micro form (according to PATANJALI and VEDIK concept body has three forms

MATERIAL body, Micro body and soul or – CHETAN body). It helps micro body to go outside the material body. With use of this VAYU the experts in YOGIK know-how, have visited remote places as well.

Hrudi prano vasenityam pano guhyamandale I
Samano nabhideshe tu udanah kanthamadhyagah I
Vyanovyapi shariretu pradhanah panchavayavyah II 30 II
(Goraksh Sanhita 30)

(Quoted from Patanjal yogpradeep, SAMADHIPAD, page 299).

Other five are:

1. NAG VAYU – Sneezing, and Impatiently spoken words etc.
2. KOORM VAYU – Bashfulness, Hesitation, doubtful mental state, etc.
3. KRUKAR VAYU– thirst, hunger, appetite.
4. DEVDATTA VAYU – sleep, Yawning etc.
5. DHANANJAY VAYU – Nutrition, Nourishment is work done by this PRAN, active even after death, responsible for nourishment to carnivores or carnivore animals.

In PRANAYAM supported by certain sitting postures of body as YOG ASANAS, sense enjoyment and desires can be controlled for spiritual enlightenment and liberation. Not only this but one can increase duration of life for perfection of spiritual advancement and liberation. Certain amount of air is inhaled in normal acts of breathing. If breathing act is forced will fully for longer period, the amount of air inhaled will be more. By holding breath at the end of inspiration and expiration the assimilation of inhaled air is also enhanced from lung alveoli to circulating blood. On the contrary if inhalation is stopped by strangling for over three minutes or more, death is unavoidable. Similarly, more air by forced inhalation will certainly be responsible for stimulating better rejuvenation of the body. By performing such practice, one immediately advances towards the liberation stage and eternal supreme atmosphere or merge into BRAHMA or becomes BRAHMA himself (GITA Chapter 14 stanza 26).

Once liberated and knowledgeable about supreme soul, one burns all material activities with fire of knowledge like any fire which turns all firewood to ashes. With controlled senses, readyness and faith in acquiring knowledge, automatically one gets knowledge. On gaining knowledge attaining eternal peace and supreme spiritual abode becomes easy. Ignorant and faith less individuals doubting supreme spiritual soul and transcendental consciousness, there is no happiness.

With the help of UDAN VAYU, it is possible that, in the highest phase of consciousness, Micro body (or SUKSHMA SHARIR), leaving material body, and visiting anywhere in the universe, at the same time it is possible to resume, the same physical body. Even today there are people present in INDIA. Recent example of this is "Dr. Vartak" a medical professional, from PUNE, Maharashtra. This has been explained above.

Ye me matamidam nityam manutishthanti manavah I
shraddhavanto na suyanto muchyate tepi karmabhihi II 31 II
(GITA 3, 31).

Those who follow teachings faithfully, and perform duty accordingly are free from fruitive actions. In opinion of Shri KRISHNA the follower of duty and teachings faithfully, as per nature's injunctions, reiterated by Shri Krishna, without any envy or doubt, becomes free from bad effects of work done or fruits of work done. In next stanza it is explained that, on disregard to injunctions laid down by Shri Krishna, out of envy, are to be considered as without knowledge and are far away from perfection in their attempts in all endeavours. Even the greatest of great person is ignorant of self and supreme soul – the BRAHM. The follower is always freed from the clutches of camouflaging effect of nature – MAYA and not getting engrossed in sense satisfaction in material world. In stanza 35, it has been warned that, imperfect duty performed as per own nature's DHARMA assigned, is far better than perfect duty in other's DHARMA. It is better to follow the path and mode of material nature with full sense of consciousness to super soul. In other words, it can be concluded that LUST and wrath when employed in super soul consciousness with total surrender to soul consciousness, it becomes friend and not enemy.

Scientifically speaking, it is this knowledge about PRAN and its ten varieties essential for human society for success and progress in material and spiritual pursues. Because of such knowledge the Sanskrit speaking civilization was advanced in all branches of sciences. Not a single branch of science which remained untouched. The concept of PRAN is still eluding modern science. It needs research in this direction.

Another fact is emphasized in this chapter is birth after death i.e. reincarnation. While answering query of Arjun that how the Gita was instructed to 'Vaivaswan' when his birth was much earlier than Shri Krishna. Lord Krishna clarified that Arjun and Krishna are born repeatedly many times before. Lord Krishna remembers all his previous birth, but Arjun is unable to retain such memory. It is a fact that all do not retain the past memory. Science behind Krishna taking birth again and again has been explained earlier which is a particle physics. Reincarnation is discussed in later chapters.

Bahuni me vyatitani janmani tava charjuna I
tanyaham ved sarvani na tvam vettha paramtapa II 5 II
(Gita 4, 5).

Reincarnation is Latin word meaning entering flesh again. Recent scholarly research "Shape of Ancient thought by Thomas Mc Evilley". For details kindly visit:

(www.jstor.org, Journal Article, Review: By Will S Rasmussen. Philosophy East and West Vol. 56, No 1 Jan 2006 page 182-191 Published by University of Hawaii press).

Gita and Vedik literature, has always favoured theory of reincarnation. Human population of the world with different religions, cult, faith, communalism (sampradaya), are divided on this subject. Couple of centuries before the concept of SOUL was not believed. But now the realization is becoming obvious globally. What does the Bible say about re-birth. (Bible. org). There are six "BASIC theories" about after death as per human race. These are:

1. **Materialism:** Nothing survives after death. It is natural accompaniment of THEISM. Gita has narrated at many places that persons involved in materialism and sense gratification are under the camouflaging effect (MAYA-माया) of luring to materialism and greed for more gratification.
2. **Paganism:** After death GHOST survives, and it goes to dead world. (Old testament Jewish notion of SHEOL).
3. **Reincarnation:** Soul goes to another body after death. It depends upon the KARM theory.
4. **Pantheism:** Death survives is same as before death. Changeless is eternal soul (Brahm).
5. **Immortality:** Individual soul survives death but not body.
6. **Resurrection:** Revival, Restoration.

Whether one believes in reincarnation or not procedure and technique has been described how previous birth can be made knowledgeable in current birth. One can recollect his previous birth.

Samskara sakshatkaranat purvajati dnyanam || 18 ||

(VIBHUTIPAD stanza 18, of PATANJAL yog Pradeep (page 518).)

Sacraments or Sanskar on mind set are imprint from two modes:

1. Memory (in seed form) and the Grief
2. Desire (in the form of VASANA). Due to Pleasure-grief, suffering, work, fruits of deeds or action results.

These are the results of deeds of past or present KARM PHAL – fruits of deeds. These are the "CHITTA DHARM". By focusing on such chittta, YOGI is revealed with record of past chitta memory. All sound scientific but it is difficult. Even then can be attempted following the technique described in YOG SHASTRA.

Next stanza advises to concentrate on some body's "chitta dharm" to get that memory of his past incarnation.

Pratyayasya parachittadnyanam II 19 II

Sanatan Dharm believes in reincarnation. Gita mentions about many incarnations of Lord Krishn and Arjun. This belief and knowledge, Gita says that it is difficult to be achieved by "un-pious" mind set.

Because of the subject which is controversial, it needs calculated and targeted research to prove or disprove the concept.

CHAPTER FIVE

SCIENCE OF CONSCIOUSNESS
(SCIENCE OF KARM YOG – ACTION)
SCIENCE OF SANKHYA YOG
(MATERIAL SCIENCE)

Abstract: This chapter deals with KARM and SANYAS YOG together. Sankhya YOG and Karm yog have been cited to explain "liberation" (Moksha) in this chapter. In fact, Sankhya and yog philosophy of Sanskrit speaking civilization has surprised the world. **Meaning of Sankhya is, "a scientific and analytical study of virtues and defects of material world".** (Explained later). Whole world is wonder struck by this philosophy. **Eternal energy which has enlivened the whole universe becomes knowledgeable by "Sankhya" and yog philosophy.** Therefore in Sanskrit literature "YOG VAISHISHTHA" says:

Naasti Sankhya samam dnyanam Naasti yog Samam balaml

No other knowledge is powerful like SANKHYA and no other power like YOG.

It is difficult to understand the contradictory two things i.e. both Karm and sanyas together. Hence Arjun demands clarification from Shri Krishn on this subject. If a person has already renounced the fruits of work how the work can be performed? Lord Krishn convinces the pupil in the form of Arjun by the theory of NISHKAM KARM YOG. Doing deeds and then surrendering the results to GOD – super soul. Instead the Nishkam karm yog is the renunciated deeds to super soul. In such deeds the ego is absent. What a scientific statement is this! The ego is responsible for the bondage of work. Missing ego can give one the true happiness.

In stanza 3 of chapter 3, Gita Lord Krishna informs Arjun that "In olden times I have told two allegiances, (1), Sankhya yogis allegiance is with knowledge(dnyan) and (2) Karm yogis allegiance is with devotion, "BHAKTI or NISHKAAM KARM yog" (without any expectation of result from work/action)".

Lokensmin dvividha nishtha pura prokta mayanagha I
dnyanyogen sankhyanam karmayogen yoginam II 3 II

(Gita, 3, 3).

A great psycho-analytical science is involved in this. Aim of science is to explore nature investigate and use it for the betterment of life to live. An answer by Shri Krishna to a query of ARJUN who is under confusion, HE explains, in comparison of both – renunciation of work and work with devotion, the latter is more superior. Human life thrives for four aims as per VEDIK understanding: "DHARM, ARTH, KAM, and MOKSH". **The science in this chapter is to achieve true happiness in life. These are psychoanalytical suggestions.** After complete enjoyment in life, humans, barring animals, look for "ultimate aim" of life i.e. liberation. Result oriented work for sense gratification responsible for material bondage and miseries in material sense gratification. It is like mirage in desert. More you run for it more and more desire develop and at the end no satisfaction. This is ignorance under the influence of MAYA i.e. camouflaging effect of – materialism in material world. To get out of bondage of material world, the knowledge that one is not material body but spirit-soul. This is not sufficient for MOKSH or liberation of conditioned soul trapped in material body. Soul is conditioned because under the influence of MAYA – the camouflaging effect, the soul is lured for sense gratification. This has been explained in previous chapter (Chapter 1.), One who has abandoned hate and desire, is free from dualities of life i.e. pleasure and grief, gain or loss, success, or failure, etc. is happily renounced and liberated. **Analytical study of material world (Sankhya yog) and devotional service, (Bhakti yog), concludes that both the yog are equal.** Follower becomes pure soul and is free. When mind and senses are in control, the person is dear to all and everyone and everything is dear to him. One is free from entanglements. Hence Gita says:

Sarvabhutatma bhutama kurvannapi na lipyate II 7 II

(GITA. 5, 7).

Such a person while engaged in daily chores of life, is aloof from the activities, and is free from sinful effects of work, and un-attached like Lotus leaf being in water but water does not stick to it.

Lipyate na sa papena padmaptravimambhasa II 10 II

(GITA 5, 10).

KARM YOG and SANKHYA YOG is discussed in this chapter to clear the doubt of confused ARJUN on the battlefield.

Sankhyayogou prithagbalah pravadanti na panditah I
ekampyasthitah samyagubhayoravindate phalam II 4 II

(Gita 5, 4).

Persons ignorant and devoid of knowledge consider that the analytical study of material world (Sankhya Yog) and devotional attitude towards soul (Bhakti Yog) are different. While learned Pandit feel both are the same.

An interesting narration is in SHWETASHWATAR Upnishad, that body is a city of 9 gates. (2 eyes, 2 ears, 2 nostrils, 1 mouth, 1 anus, 1 genitals). The spirit is entangled and embodied in such city. The living entity in this conditioned stage, identifies with the city – the body. When one identifies due to the ignorance of oneness with the city he is conditioned and not free from bondage. When one becomes knowledgeable with soul and super soul within the body, he becomes soul and Super soul himself even when within the city/body.

Navadvarepuredehi hanso lelayate bahihi I
vashi sarvasya lokasya sthavarasya charasya cha II 18 II
(SHWETASHWATAR UPANISHAD. 3, 18).

A small fraction of super energy or superior soul, temporarily embodied in material body, is master of the current body. Even then the master is neither duty bound for action nor work, or effect of work outcome. The action and resultant reactions are performed as per natural instincts of the material body. But due to ignorance one identifies oneself with the material body, then he feels ownership of actions and its resultants. This ignorance results into bodily sufferings and distress. Ego (Aham or 'I') comes in between. This leads one to assume body's sinful or pious actions as his own. In fact, neither sinful nor pious deeds of body are by the soul. Proper knowledge and shunning of ignorance help destroying the illusion. Thus, learned PANDIT sees everything equal including BRAHM and the mind settles in self-realization and realization of BRAHM. He enjoys eternal joy PARMANAND, (sat chit aanand).

Dnyeyah sa nitya sanyasi yon a dveshti na kangkshati I
niradvandvo hi mahabaho sukham bandhatpramuchyate II 3 II
(Gita, 5, 3).

Dualities of life – hate, desires for fruits of activities, pleasure, grief etc. are to be given up. Such renounced person (Nitya sanyasi) is free from the bondages (bandhatpramuchyate) of material life, is completely liberated.

In next few stanzas Lord KRISHN explains qualities of realized person and the practice of YOG for self-realization.

In fact, the whole universe and the manifest material world is expression of super soul. In 14th chapter of GITA, it is emphatically stated that:

Mam yonirmahadabrahma tasmingarbha dadhamyaham I
sambhavah sarvabhutanam tato bhavati bharat II 3 II

(GITA 14, 3).

It is the BRAHM which is the source of all material world. (details in chapter10 of this book). The super soul is responsible for impregnation of this brahm. That is the way the whole living and nonliving material world manifests (Brhdaranyak upnishad ch. 1, Brhman 4, stanzas 1-7 referred above. Page 133). Therefore, whatever one does be dedicated to super soul. Obviously, this act will free the person from the good or bad effects of the actions performed. Keep on nature's destined work with dedication to super soul. Accordingly, with such "mind set" father of creator of atom bomb (Oppenheimer) got solace from bad effects of destruction device. This is a glaring example.

(The GITA of J. Robert Oppenheimer, by James A. Hijiya, Professor of History, University of Massachusetts, Dartmouth. Proceedings of the American to Philosophical Society Vol 144, No 2, June, 2000).

Oppenheimer knew GITA thoroughly, and studied other Sanskrit texts. He knew quite a bit about MAHABHARAT. Oppenheimer believed that he, being a physicist, specifically a nuclear physicist, he is duty bound to perform duty as a scientist. He knew that the Manhattan project of USA, is going to deliver a dangerous and lethal nuclear weaponry, at the same time as scientist, had an obligation to serve on the project.

In the same context Gita is clear about physical and spiritual knowledge. A comparative statement between VIDYA and AVIDYA is in the stanza quoted on page 80 chapter 2. Knowledge of both physical sciences (avidya) and spiritual science (Vidya) are essential to be happy and liberated. (see details on that page – from ISHAVASYOPNISHAD). Because of this philosophy physical sciences were also highly developed.

Students from all over the world used to study in university of TAKSHASHILA (and NALANDA), WORLD'S first university with residential facility. Foreign students used to join these universities especially for medical education, which was highly developed during Vedik period. (see Charak Samhita – Medicine and Sushrut Sanhita – Surgery, having 8 branches of Aayurved-Ashtang Aayurved) Details are out of the scope of this book of exploration of science in GITA. The hidden science in these Yoga shastra, described above is explained below:

Gita in stanza 14 of chapter 5 says:

Na kartrutvam na karmani lokasya srujati prabhuh I

na karmaphalasanyogam swabhavastu pravarjtate II14 II

(Gita 5, 14)).

Reality is the Super soul neither create activity nor action and the results thereupon. All depends upon the mode or SVABHAV. Every action is NATURE dependent and not on GOD. Nature is created by GOD. The rest is on nature dependent automation. Every action is carried out through Nature or PRAKRITI. Supreme soul does not allot sinful or pious activity to anyone. Ignorance and affection of individual and the influence of camouflaging, luring of nature (माया) which hides the real knowledge. The following stanza explains this more emphatically:

Na datte kasyachitpapam na cheiva sukrutam vibhuh I

adnyanenavrutam dnyanam tena muhyanti jantavah II 15 II

(Gita 5,15).

Super soul (Vibhuh) does not accept any body's sin or pious deeds, ignorant covered knowledge attracts living beings.

This stanza has wide repercussions on our behaviour. God does everything is a wrong conception and must be shun off if one believes in GITA preaching. Our existence is expressed through nature (Prakriti). Without dependence on GOD, one must do needful action wherever and whenever essential. This is the scientific explanation for the subject of BEHAVIORAL SCIENCE. When one is enlightened with knowledge, everything gets revealed. True knowledge can only bestow the real wisdom. Lord Krishn informs ARJUN (through Arjun and all BHARTIYA and all humans) to fight against injustice.

Work areas or sphere of work and behaviour of SANKHYA YOGIS in material world of nature (Prakriti), is explained in Gita chapter 3 stanza 28:

Tatvavittu mahabhavo gunakarmvibhagayoh I
gunaguneshu vartant itimatva na sajjate II 28 II

(Gita 3, 28).

The meaning and science hidden in this stanza is explained below.

Elemental knowledge about different virtues of materials interact and engage (guna guneshu vartanta) among themselves (is absolute truth-tatvavittu) and never get attached is my opinion.

Meaning of this stanza is taken by Spiritual masters is religious. Spiritual meaning can be knower of absolute truth, Oh Mighty armed Arjun, does not indulge or get attached in sense gratifications and fruitive results.

Scientifically the explanation is as follows:

To understand the above stanza, the quality particles of DNA of cell need to be understood. After the research of Dr. Har Govind KHURANA (Nobel prize in Physiology, in 1968 shared with R W Nirenberg, for genetic code and its function in protein synthesis) and the theory of quality particle arrangement of DNA in the cell, which decide the heredity and genetic constitution of an individual. Bodies of "Multicellular organisms" are made of numerous cells as building blocks. Each cell has a nucleus. In the nucleus there are chromosomes. The chromosomes harbour genes. These genes decide the heredity of the possessor. Genes are in the form of coils. The genes have the quality particles in serpentine coils. Serpentine arrangement of quality particles in coils preserve their place for generations. **But are subject to change gradually after a long lapse of time. This depends upon the**

cultivations acceptable and accepted by the person. Sanskrit speaking civilization was aware of this scientific fact hence put forth the theory of Imparting of "SANSKAR" to progeny as well as Gotra matching hypothesis in VIVAH SANSKAR (marriages). Genetic change is possible with considerable cultivations accepted by the individuals. Here the cultivation is important. The quality of the person and the SPECIES gradually and slowly change to form new quality species. The quality particles in DNA, are responsible to decide the quality of behaviour of the person, like greed, anger, envy, affection and so on. These particles do not have any effect on the Jeevatma or entraped soul in material body. But JEEVATMA feels, he is the same as material body. It is the effect, of continuous cultivations accepted in mind-set (CHITTA). In the stage of the SAMADHI the qualities get subsided. Thus, quality improves. The word GUNVIBHAG and KARMAVIBHAG word of the above stanza means virtues and action parts about arrangement of the quality particles in serpentine coils of chromosomes of DNA. It can be compared with electricity flowing through the straight wire and coiled wire. Coiled wire shows magnet effect when electricity flows through but not in straight wire. YOGI gradually develops all good qualities with continuous practice of meditative sessions. YOGI never gets change instantly by one or two meditations. It is gradual and a slow process. It is called as the awakening of SERPENTINE KUNDALINI in MULADHAR CHAKRA (see above – serpentine coil of gene). Chromosomes are visible under a microscope during mitosis (cell division) because of the short, fat appearance after tightly coiling. Walther Flemming, a German scientist, was the first man to observe chromosomes while studying the division of salamander larvae cells in 1882.

But Sankhya Philosophy about material analytical science is different. Sankhya is analytical study of material world, as described earlier. Through the SANKHYA philosophy the spirituality is awakened. This point must be remembered. Analytical study of material world is clearly described by Maharishi KANAAD in "VAISHESHIK SUTRA".

Meaning of Sankhya is, "a scientific and analytical study of defects and virtues of material world". Ponder on highlighted words in stanza below:

Doshanamcha gunanamch pramanam pravibhagatah I
kanchidarthambhipretya sa sankhyetyupdharyatam
(Mahabharat)

To understand meaning of – Virtues interact, Gita 3. 28, one must unravel the knowledge of principles of Maharishi Kanaad in **VAISHESHIK DARSHAN and NYAY DARSHAN** by Rishi Gautam. Vaisheshik Darshan is also known as the"KANAAD or AULUKYA". Because of special imaginative work it is nick named as special or VAISHESHIK. Because of Kanad Rishi and his father Uluk Rishi, it is also known as KANAAD and OULUKYA. At some place the gotra (paternal lineage) of Kanaad is described as KASHYAP. Matter specific generated material (dravya), Virtue (gun) work (karma), general and specific study of elements gives knowledge of elements (tatva dnyan). Nine types of elements described in Vaisheshik as follows:

Pruthivyapstejovayurakasham kaalo digatma mana iti dravyani II5II
(Vaisheshik 1, 1, 5).

Split highlighted word is = Earth (Pruthvi), Water, Fire (tejo) Air, Space (aakasham\) Time, Direction and Soul, Mind, That's all are elements (iti dravyani)

Nine elememts (DRAVYA) are Earth, Water, Fire, Air, Space/ Aakash, Time, Direction, soul, and Mind. For details one may read original work. Virtues and subjects of these first five are – Smell, taste, liquid or juicy, Appearance, Touch, sound or word. Sense organs for these senses are Nose, tongue, eyes, skin, ear.

Work or actions of elements or Karm are of five types.

Utkshepanam vakshepanam aakunchanam prasaranam
gamanamiti karmani II 7 II
(Vaisheshik 1, 1, 7)

Utkshepan = Throw upwards. aakunchan = Contraction, avakshepan = Fall down, prasaran = Expansion, Gaman = Circulation or other different work. These are present in all elements. These are described **DHARM of elements** (and not religion as per modern understanding). It can be noted here that all this is possible when elements have particle in it or made up of minute particles. (This is Kanad's "particle Physics" as per modern science. For details kindly refer to original work). *(Patanjal yog prdeep, code 47, Gita press Gorakhpur, page 48 – 54).*

NYAY DARSHAN. "Nyay sutra" is written by RISHI GAUTAM. It deals with the evidence and is evidence based. **Taking help of evidences examination of elements is Nyay.** Evidence can be direct (Pratyaksha) or indirect (आगम). Details of this Nyay darshan by Gautam Rishi, are beyond the scope of this book. Reader may visit original work for details. Elements are analysed and examined and studied with the help of evidence. Different **names of evidences** are mentioned in the following stanza:

Praman prameya sanshaya prayojan drishtanta sidhantavayavatarka nirnayavadajlpavitanda hetva bhasa chhala jatinigrahsthanam tatva dnyanannihi shreyasadhigamah II 1 II

(न्यायं I, I.).

Proof. (Praman), Hypothesis to be proved (Prameya), Doubt (Sanshay) Purpose (Prayojan), Illustration (Drashtant), Theory or principle (Sindhant), Componets (Avayav), Arguments. (Tark), Decision (Nirnay), Debate (Vaad), Dispute (Jalp), Controversy (Vitanda), Fallacy (Hetvabhas), Deception, Stealth (Chhal), Species (Jaati) and Place of control (Nigrahsthan). The knowledge of all these one gets liberation. That is freedom of understanding the original worth of the elements.

Here one thing be noted is that the ultimate aim, of human life is taken as for the liberation of the practitioner; that is liberation or MOKSH. Knowledge of PRAMEY one can be liberated (Moksha). Knowledge of प्रमाण etc. one gets knowledge about elements also. This does not grant one not to think about science involved in the

analytical scientific evaluation of elements. Unfortunately, modern commentators or expositions of GITA took the description as religious and spiritual only scientific aspect was totally ignored (see Vidya and avidya above-chapter 2 page 80). Modern commentators on Gita quote references from western authors. This confusion may be the influence of such impact of thought.

According to NYAY DARSHAN there are 4 main evidences, Direct (Pratyaksha), indirect (anuman), similarity (upaman), prophecy (aagam).

Because of this knowledge in **common people** Gita cited this as examples in the Stanzas. Chapter 3, shlok 3.; Chapter 5, shlok 4-6, 8-12, 28. No details given. Gita states Sankhya as DNYAN YOG and SANYAS YOG. Sankhya and yog are oldest philosophies of Vedik India. Hence it needs to be explained from other sources of Sanskrit literature. Super soul who is cause of whole manifest and unmanifest Universe can only be known and understood by SANKHYA and YOG. Yog can be mastered by controlling mind-set (CHITTA) while Sankhya by proper knowledge. Gita 3, 3. The Sankhya is knowledge (DNYAN YOG) and work/action without any attachment is KARM YOG.

> *Lokesmin dvividha nishtha pura prokta mayanagha I*
> *dnyan yogen sankhyanam karma yogen yoginam II 3 II*
> *(Gita 3, 3).*

In Mahabharat: It is mentioned that study of virtues and defects in any substance is known as "SANKHYA":

> *Doshanam cha gunanam cha pramanam pravibhagath I*
> *kanchidarthambhipretya sa sankhetyupdharyatam II*
> *(mahabharat).*

Originator, Promoter and Propogater of Sankhya is KAPIL MUNI. He was an expert muni hence Gita says Sidhanamcha Kapilomuni. (Gita 10, 26).

After establishing existence of Aatma and Paramaatma (VIDYA) with study and practice of (SADHANA) of YOG, the philosophers also analysed existence and composition of different elements of material world (AVIDYA). Maharishi KANAAD and GAUTAM both after discussing spirituality knowledge also advised study, analysis of elements in their originality by YOG-SADHANA. Following statements from VAISHESHIK DARSHAN authenticate above statement (VAISHESHIK 9. 1. 11-14):

Aatmanyatmamanasoh sanyogavisheshadatmapratyaksham II 11 II

Special unification (sanyogvishesh) of ATMA and mind (aatmanyatman) by YOG-SAMADHI, reveals soul originality (datmapratyaksham\).

Tatha dravyantareshu pratyaksham II 12 II

And elements's originality also revealed (by yog samadhi).

Asamahitantahkarana upasanhrutasamadhayastesham cha II 13 II

Those who have attained samadhi, get revealed the elements without samadhi also.

Tatsamvayat karmaguneshu II 14 II

similarly, when involved in those elements (tatsamvayat), virtues and actions (karmaguneshu) of elements also get revealed to them.

As AATMATATVA is revealed by SAMADHI and YOG SADHANA, similarly elements also get revealed. Liberated person (Samadhi Prapt) the elements beyond senses also get revealed to them. Work, virtues of those elements get exposed due to their involvement into the elements. Real nature and essence (absolute truth) of elements (tatvadnyan) get exposed to them. Similar expression is in Nyay darshan.

Vedant is basically end of knowledge i.e. beyond which nothing remains to be known. Sankhya and Yog are the means to know the Vedant. In other words the these are the means of knowing the elementary knowledge of the material manifestation also. Mundakopnishad in Chapter 3, Section 2, in stanza 6 and 7 have narrated very beautifully this scientific fact.

Vedantavidnyan sunishcharthah sanyas yogat yatayah
shuddhasatvah I
te brahmalokeshu parantakale paramrutah
parimuchyanti sarve II6II
(Mundakopnishad 3, 2, 6).

Those who have understood perfectly the ultimate scientific knowledge of VEDANT and with renunciation attained pure mind-set, certainly get freedom of understanding materially and liberated spiritually. This emphasizes a fact that Ved and Gita advocate getting knowledge is a form of liberation.

Gatah kalah panchdasha pratishtha devashcha sarve pratidevtasu I
kamarani vidnyanamyashchya aatma parevyaye
sarva ekibhavanti II 7 II
(Mundakopnishad. 3, 2, 7).

All 15 (5-senses, 5-action senses, 5-panch mahabhut), unite with their respective source. Work actions and scientific aspect of soul, all unite with super soul.

Those who have definite special scientific knowledge of Vedant and purified their conscience (Antahkaran) at the end become brahma themselves (stanza 6 above). It is further explained in next stanza above (7) that the body consisting of 15 phases – 5 basic elements (panchmahabhut), 5 work organs – Karmendriya, and 5 sense organs – Dnyanendriya). All merge along with their respective 'devta' i.e. basic elements (Earth, water, tej, air and aakash) and their 5 Subjects – form, juice, smell, touch and words (roop, ras, gandh, sparsh and shabda) along with their sense organs (eyes, toungue, nose, skin,and ear). Other 5 being karmendriya work organs – hand, leg, talk, anus and penis).

All 15 KALA give us knowledge about the soul. Energy or DEV inside body merge with energy outside body at the time of death (**sarvaekibhavanti** II 7 II).

This is what Gita says in Ch.5 Stanza 24:

Sa yogi brahmanirvanam brahma nirvanam brahm
bhutodhigachyati II 24 II

(Gita. 5, 24).

A perfect yogi is happy within, highly active mentally within, rejoicing within (Sat-chit-aanand) and illumined within. He is liberated in fact he is BRAHM himself.

CHAPTER SIX

SCIENCE OF CONCENTRATION (AATMASANYAM YOG) DIETETICS NEUROANATOMY SCIENCE OF DHYAN AND MEDITATION

Abstract: In this chapter a great stress is laid on MIND and mind-set. Concentration depends a lot on mind. The first and foremost essential condition for concentration is "self-restraint". What is self-restraint? It is easy to advise than practice. First four stanzas, of this chapter stresses, on preliminary need, method and technique for meditation, concentration etc. The ATM-SANYAM YOG is the basic theme of 6th chapter. It is the practical side of yog. Control on self provides tremendous strength to the yogi. Yogi gets 'clair' powers by regular practice – DHYAN and meditation. The need for concentration, is **determined mind-set (manah Sankalp)** and restrained desires for sense gratification.

Vedik civilization realized the importance of mind over body. Mind has total control over body. Mind controls reflex activity of nervous system. For example, if mosquito bites sitting over skin, reflex neuromuscular action will remove the mosquito instantly. But if mind decides not to do so reflex can be suppressed. Similarly, controlled cultivation of

SANSKAAR on mind gradually change the behavioural attitude of the person. With such a cultivation effect on mind changes the physical material body. This change slowly and gradually changes the genetic quality particle arrangement of the DNA of the cell and in turn the attitude of the body (details discussed later in this chapter). Sage Vasishtha (Illumined Sage, Teacher for Young Ram. He taught Young Ram for 18 days. Book on this is Yog vasishtha) discussed about mind in YOGVASISHTHA.

Mind is basically a process of imagination. It is unsubstantial. Its existence is imaginary. yet it thrives due to consciousness. In Yogvasishtha it is very nicely mentioned:

Chitchhakteh spandashakteshcha sanbandhah kalpyate manah I
mitthyeiva tatsamutpannam taduchyate II 88 II
(YOGVASISHTHA 13, 88).

Knowing power (chitshakteh-chitta+shakti) and pulsating or vibrating power together create imaginative mind. It is unsubstantial illusion or existence is false. The mind gives birth to thought is itself imaginary. But entire material body is possessed by mind! Interestingly thoughtful mind can cut the evil mind as an axe (***kutharene***) can cut a tree:

Manaseiva manechhitva kutharene padpam I
padam pavanmasadaya sadya eva sthiro bhava II 35 II
(YOGVASISHTHA 13, 35).

Cut mind, using mind as axe – like cutting a tree. Step into self within and remain unshakable.

Renunciation of fruits of action and work is first step towards concentration. This does not mean work and action to be given up and no duty to be performed. Only KARMPHALA to be given up mentally. 5th stanza speaks about mind. Unscientific presumptions and faith that one cannot elevate oneself unless some supernatural being, sage or saint help the person. Gita says in a different to this in 5th and 6th stanza.

Udharedatmanatmaanam natmanamvasadayet I
aatmeiva hyatmano bandhuratmeiva ripuratmanah II 5 II

(Gita 6, 5).

Mind is friend and if controlled and conditioned otherwise mind is enemy. Controlled mind prevents the person from being swayed away by glitter and attraction of material world due to camouflaged effect or due to MAYA.

Bandhuratmatmanastasya yenotmeivatmna jitah I
anatmanastu shatrutve vartetatmeiva shatruvat II 6 II

(Gita 6, 6).

Controlled mind is like brother otherwise mind is the worst enemy.

Absolute truth is same everywhere. Supposition or faith cannot become truth. A person is capable of elevating and degrading himself, by his intellectual or mental capability. What a scientific statement is in 8th stanza:

Dnyanvidnyantruptatma kutastho vijitendriyah I
yukta ityuchyate yogi samaloshtashmakanchanah II 8 II

(Gita 6, 8).

Wisdom and scientific mental bend as well as controlled senses, in a yogi hardly see or feels no difference in STONE or GOLD. Because these are the stages of mind-set about absolute truth.

What is MIND? Where it fits anatomically in body? Which organ is its abode? How physiologically it can be described? What Sanskrit literature knows about the mind?

The word mind is intriguing. It is a product of living body. Play of mind is inevitable in a living body. Body's animation is due to presence of mind inside the body. In fact, it is the life support of body. Hence in Sanskrit it is the JEEV (a living entity). Yog Vasishtha Ramayan narrates that the mind is the seat of awareness of soul. As per the function of mind there it is described variously. It is recognized by different names, Jeev, ego, buddhi – intellect, mind, prakriti – nature, body, and self.

Jadajad drishormadhyam yattatvam parmatmikam I
tadetdev nanatvam nanasangyabhiratatam II 56 II

(YOGVASISHTHA 71, 56).

Display of Nature is the whole manifest world. Even though it is mysterious, but it is revealed to the spiritual knower who knows the inner self. The power, of the body and its parts, are due to the relation and mastery of Mind. Body with all its parts, is gross, while mind is subtle without any parts. Its existence is internal and invisible. It is the product of the body and its parts and exists in three states – wakefulness, sleep and dream.

In stanzas 11 – 14 Gita explains the practical posture, positioning of spine and body at a particular place and concentration of mind and mind-set at the tip of nose or root of nose between two eyebrows (It is described later on at appropriate time).

What is the science in this site, positioning and body posture etc.? There is external illusionary energy around material body. This camouflaging effect (MAYA) of the universe distracts mind and mind-set lures towards the sense gratification and sense enjoyment all the time. Thus, mind is misled and distracted away from concentration and spirituality. The technique described by Gita distraction is controlled and concentration on the object of mind-set gets stabilized. Stillness of mind and concentration gets enhanced. "Still mind" is more active and consumes less energy as explained earlier scientifically. Once in SAMADHI spiritual and material knowledge gets revealed to the practitioner (SADHAK) – see chapter 5.

As a helping hand for yog and meditation an appropriate diet has been stressed. What diet is good for better yield and achievement using meditation and practicing yog? Diet plays a great role in life. In medical practice appropriate diet is prescribed to enhance curative effect of medicines, similarly, diet influences mind set. **Science of diet** was known to VEDIK people. A lot of attention about diet, has been given in modern science. Not only quality of diet but stress has been on the quantity of diet as well. In Gita Stanza 16:

Natyshnatastu yogosti na cheikantmanashnatah I
na chatisvapnashilasya jagrato neiva charjuna II 16 II
(Gita. 6. 16).

It laid a great importance on eating too little or eating too much, both are harmful and disturb meditation and concentration. In other words, GITA has played a role of a dietician for health, meditation and yogik concentration in the form of Lord KRISHN. Dietician plays a great role in modern medical science. Diet – includes type and quality of food. Sages and Rishis have considered factors like sleep, dreams, work actions etc. which influence mind-set, meditation and yog, apart from diet. Too much of sleep and too little sleep also interferes in meditation and YOGIK achievements. In 17th stanza of 6th chapter Gita states:

Yukta aahar viharasya yukta cheshtsya karmasujaha I
yuktasvpnav bodhasya yogo bhavati duhkhaha II 17 II
(Gita 6, 17).

GITA stanzas have played a role of a dietician as per the modern science is concerned. The material body demands always eating, sleeping, defending, and mating. All these and overindulging any of these, disturb advancement of YOGIK practice. Appropriate diet, sleep and work keeps one alert and concentration towards soul/super soul will always help in advancement to transcendental meditation. At the same time scientific development, research and analysis which needs concentration, and meditation on the concerned topic, is also helped by proper diet. Diet disciplines mental faculties was well known to the Sanskrit speaking civilization and the contemporary society. Regulation in one's work, sleep, wakefulness, speech, and all other bodily activities is important for the practice of yoga. In fact, YOGI burnout all, natural or bodily demands materialistically, mentally, or physically. With un-appropriate diet, mind becomes restless, turbulent, and very, strong. It is more difficult than the controlling of wind. By practice, determination and detachment YOGI can curb it. Appropriate quality and quantity, of diet, definitely help in curbing the restless mind. Devoid of all material desires the establishment in YOG is firm and firmer. As per quality of food ingredients, "Diet" can be divided into three groups as per VEDIK literature.

1. SATWIK (PIOUS, enhance energy),
2. RAJSIK, (PRINCELY, food that increase energy) or
3. TAMASIK (energy for Sinful, Ruthless, merciless attitude).

The three virtues of diet as above, needs explanation from Vedik or Yogik literature.

> Satva gun has happy form of life,
> Rajo gun has unhappy form of life while
> Tamo gun has fascination and affection form of life.

The strength of these attributes: are SATVA GUN = enlightenment, RAJO GUN = motivation and TAMO GUN = restrictions. Therefore, behavioural description of these three are:

> SATO GUN = peaceful enlightened behaviour,
> RAJO GUN = aggressive behaviouur, and
> TAMO GUN = restriction behavior.

Satvam laghuprakashkamishtamupashtambhakam chalam ch rajahl guruvaranakmeva tamah pradipavchcharthato vritti
(sankhya karika 13)

(Adopted from Patanjal yog Pradeep, ch 4, page 110.)

DHARM of these, three GUN are:

SATVIK food diet is based on Vedik medicinal – Aayurvedik and transcendental yogik suggestions and principles. It is nourishing and rich in nourishing values, and virtues. The diet is meant to include pure, natural, essential clean eating practice. Mind remains calm and enlightened. Person becomes humble. This provide pure energy which flows throughout the body. The food is unprocessed. This diet keeps body clean and healthy for keeping mind peaceful and happy. Food items included, are Milk, Curd, Green vegetables, Nuts, Honey, Butter, Paneer etc.

RAJASIK food is spicy, sour, pungent and hard to digest. Such food gives elation and excitement, stimulation. This food makes the person sensual, passionate. This type of food can also cause stress and diseases. Tea, coffee, cold drinks, Garlic, onions, Live-stock etc. are all rajasik type.

Tamsik food is Meat, eggs, **stale food**, Effect of such food is person gets anger, he is revengeful, full of Laziness with no intention to work, and tension etc.

Effects of these three GUN-DHARM:

An interesting description exists in PATANJAL YOG SHASTRA. All the three GUN remaining in their separate form, and still the effect result is called as analogous (SADRISHYA) or SAMYA, (resemblance) form and responsible for the formation of earthly elements in the universe. These three GUNAS, when suppress each other the resultant is adverse (VIDRUP effect). It is known as contrasted effect VISHAM. Basic elements (23 elements) including all five basic elements (earth, water, air, light, space) and contrasted effect of GUN the manifest nature is created (like Chromosomes of DNA and particle arrangement in coils of chromosomes). All this is revealed in YOG SAMADHI. Pleasure-grief, Desire, Knowledge, Efforts, malice/grudge, all these are DHARM of "Mind – set", Conscience as per SANKHYA YOG. The mind set and soul are separate from each other. Soul is CHETAN TATVA-knowledgeable. Because of union of these GUN, hence associated with the soul. These GUN only, are responsible for the creation of nature, Trident root elements and soul both are eternal. The contiguous union of both is also eternal (beyond time scale). Living beings Animals and man need diet immediately after birth. Breast feed to start with and then food from material world. This is essential for energy, immunity, and sustenance of life. Animals keep on eating all the time and go on storing in the form of fat. When food is unavailable stored food is consumed till fresh food is made available. But humans are different in food habit. Food is available and stored for daily consumption. The eating is 3 to 4 times a day. Food is made available all the time. But animal instinct of storing in the form of fat continues. This results into unhealthy bodies. The effect may be to acquire diseases (-Metabolic or infectious etc.). But in YOG and Meditation one needs control on MIND and Body. Diet influences not only body but MIND-SET as well. It is said that man behaves according to the diet-food he eats. This knowledge was existing with Sanskrit speaking civilization. The diet was based on AYURVEDIK principles. The diet was divided into SATVIK, RAJASIK, TAMSIK. This has been described earlier. Not only **balanced diet but quantity of food** is also specified. Described earlier. Type of food and quantity of food is prescribed in diet.

Susnigdhamadhuraharshchaturthanshvivarjitah I
bhujyate shivasamprityei mitaharah sa uchchyate II 58 II
(HATHPRADEEPIKA. I. 58).

Not only this how much quantity of food be consumed has been impressively described in above stanza: One fourth of space (highlighted and underlined word in above stanza), for water and air to play in stomach bag. {In stomach bag two parts food, one part, water and one part, for air to circulate}. (Details in Aayurvedik literature on diet). Type of food consumed decides the character and Nature of the person. Bend and behaviour of nature of the person is as per the type of food consumed. It has been discussed earlier.

Not only quality and quantity of food but a great stress has been levied on how the food is earned and prepared:

Mind has a great role in controlling all bodily actions. Sense gratification is a 'natural-instinct' of all living beings. If mind can be wilfully controlled, it gives miraculous results. Diet has a great effect on mind and body. **Diet/food must be very carefully selected. Food must be pious and piously earned. There should be no cruelty involved in getting food. Probably because of this vegetarian diet was advocated. Food cooking and preparation must be with hygienic conditions and with pious mind-set of the cook.** With diet and mind control one can steadily progress towards the transcendental meditation and can achieve a lot physically, spiritually, and philosophically at the same time can advance scientific persuasions. If mind is set free, it becomes enemy, thus Gita says:

Udharedatm manatmanam natmanam vasadayet I
aatmeiva hyatmano bandhuratmeiv ripuratmanah II 5 II

(Gita 6, 5).

A lot of stress has been given on MIND in this chapter. One can realize by the fact that all activities of the material body depends upon mind. Even when somebody is speaking near you but if your mind is not focused on listening, what the person is saying the listener is unable to listen the talk or understand it. Similarly, if mind is not on what one sees the vision becomes zero i.e. the person is unable to perceive the object in vision. Therefore, control of quivering mind is essential for whatever the action performed by the physical body. In material world the material body is under the control of mind was essentially realized by the Sanskrit Speaking civilization. The behaviour of mind depends on diet a person consumes. By controlling mind highest ever knowledge was achieved by that civilization. All branches of sciences were successfully advanced. Most difficult being spiritual science, planetary science in space, and calculation of movements of Stars, Planets, galaxies etc. were successfully understood and calculated. Not only this but the effect of these stars on the material body, was also studied, learned and understood. Thus astrology, astronomy, palmistry, branches were developed. Stanza 6 reiterates the mind effect on material body:

Bandhuratmatmanastasya yenatmeivatmana jitah I
anatmnastu shatrutve vartetmeiva shatruvat II 6 II

(Gita 6, 6).

Meaning that if mind is won it is the best friend and if one fails to win the mind, the mind becomes the worst enemy for all the life and thus he loses all the battles of life. In other word, uncontrolled mind controls the person's material body and mind.

Attention focused on single thought, controlling mind, senses, actions, and all activities, practice of YOG clarifies the heart. (Gita 6, 12).

Tatreikagram manah krutva yatchittendriyakriyah I
upavishyasane yunjyadayogma tmavishudhaye II 12 II

(Gita, 6, 12).

In subsequent stanzas the technique of sitting posture, position of head, neck and spine in one straight line, focusing eyes at a point in between eyebrows, described for subduing mind easily.

Samam kaay shirogrivam dharayannachalam sthirah I
samprekshya nasikagram swadisha shchanavalokyanam II 13 II

(Gita 6. 13).

This is further helped by following the principles and practicing of BRAHMACHARYA. It is YOGIK lifestyle for student of any discipline and seekers of absolute truth. The same view is expressed in stanzas of GITA. BRAHMACHARYA is when a person controls his CHITTA (mind-set) through ascetic means. A monk like abstinence or celibacy along with study of literature like VEDAS, study of BRAHM or absolute truth (Gita 6. 14). This is the scientific method described in GITA for achieving any knowledge in the universe. The ancient technology developed and propagated since time immemorial and narrated in the form of GITA to ARJUN.

Lord Krishn has explained practice of yog. The benefits, techniques with where and in what position to practice yog. What diet is helpful to achieve success in practice of yog. How a successful yogi looks and behaves. What if transcendental attempt is incomplete or is not a success. What is the outcome of such an individual? Let us understand the scientific aspect of the yog emphasized by Lord Krishn. Details are available in yogshastra.

Every living being is in some form or other is involved in getting rid of grief and obstacles in living happy life. Grief can be as per fate (Aadidaivik), physical (Aadibhoutik), spiritual (Aadhyatmik). Considering this, three main facts needs to be pondered.

Soul (Chetan tatva) who suffers grief, what is its form? Nature is basic element of material body. The basic element of soul is not grief. Its relief is mandatory. By understanding soul and its real form one can get rid of grief. Nature is basic element of manifest material body. Detailed knowledge of the soul and body may obviate grief. Super soul (Paramatma tatva). Apart from soul and body there is another element which in fraction as soul appears in material body of living beings. By reaching to such knowledge grief can totally be obviated. Union with super soul or YOG was developed by sages of Sanskrit speaking civilization. Mind is flickering and unsteady. It needs to be controlled (Gita 6, 26). Shri Krishna has detailed, the place and posture for yog practice:

Shuchou deshe pratishthapya sthiramasanmatmanah I
natyuchchhitam natineecham cheilajinkushottaram II 11 II

Tatreikagram manah krutva yatchitendriyakriyah I
upavishyasane yunjad yogmatma vishuddhaye II 12 II

In a secluded and sanctified seat or "Aasan" of dear skin or soft cloth on a grass (KUSHA), Seat not too high or low, sitting attentively with controlled mind-set, senses, and activities, one has to practice yog. In stanza 13, 14 Gita has prescribed sitting posture with celibacy, concentrating on one point one should meditate. The place, position and posture of body is highly scientific for the ultimate aim, for practice of yog. Here the aim is to

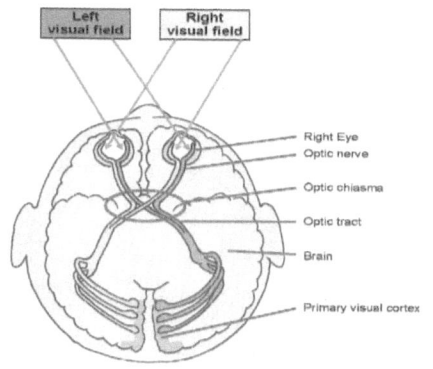

sustain attentiveness of mind set. This is all because meditation is not possible while standing, running, or lying position for obvious reasons. This posture makes one for easy concentration on desired object. The focus of concentration is advised at the tip of the nose or in between the two eyebrows.

Scientific fact about this site is:

Right and left optic nerve bundles come out of Optic chiasma and reach two eyeballs

A third branch is blind end (this is only yogik belief) reaches between two eyebrows to be confirmed anatomically.

Bhagvan Shankar, the originator of YOG SHASTRA, has been depicted as having three eyes. The third eye is in the centre of two eyebrows. Third eye is for DIVYA DRISHTI or YOG DRISHTI – divine sight. Concentrating on this site a person can develop DIVYA DRISHTI. Yog shastra believes that, the third branch of optic nerve has blind end and reaches this point. Yogi believe in concentration on this point and can activate this nerve for YOGIC VISION. It is a point for targeted research as per modern scientific standards. In Sanskrit literature also this site is advocated for concentration.

Possible scientific explanation can be as follows. There are association bundles in brain cortex (short, long and 'U' shaped). The long bundles connect occipital lobe region (visual cortex) to inferior pole of frontal lobe, which is situated in between two eye brow region. The connecting bundles are association of all senses from various brain centers.

These bundles correlate brain functioning

Inferior pole of frontal lobe   Occipital Lobe-Visual centre

Diagram of association bundles. (Long, short, and U-shaped bundles)

(https://en.wikipedia.org/wiki/Association_fiber) Inferior pole of Frontal lobe

Visual centre in hind brain in association with frontal lobe (inferior pole) develops mental picture. Association fibre bundles from other specialised centre of brain develop communication with frontal lobe helps in mental picture development.

(The Telencephalon. D.E. Haines, G.A. Mihailoff, in Fundamental Neuroscience for Basic and Clinical Applications (Fifth Edition), 2018)

Time of meditation is also important. Maximum silent surroundings as per natural laws is "BRAHM MUHURTA". This time is 3 AM till before sunrise. Brain is active at its highest, in silent dynamism. Brain, mindset work without resistance, with minimum energy consumption, and maximum yield. It can be compared with uniform speed in a straight line without ups and downs and turns in speed. Body, spinal cord, brain stem and the brain fall in one line. It is easy for the coiled energy (Kundalini Shakti) in sacral bone in MULADHAR CHAKRA to get unfolded and rise to top of the head where the BRAHMARANDHRA (place of anterior fontanelle) is situated. From here the union of raised KUNDALINI with super soul is easy. This is biologically and scientifically sound.

DHYAN. Mind and mind-set are always lured by camouflaging effect of (MAYA) of material world. Sense enjoyment and sense gratification becomes primary desire. Tendency of mind is to wander. Concentration of mind at a desired or decided subject need efforts in controlling mind. It is possible to control and regulate mind wilfully. This is by meditation or Dhyan. Attempts to focus mind as described by Gita as mentioned above. Such oneness of mind-set with desired subject of action is called DHYAN. It is meditation. But mind has tendency to wander again and again in meditation. So, one must try hard to concentrate again and again repeatedly. This continues till the mind is in total control and no more travelling of mind here and there.

Every living entity in this material world accumulates result and reactions of work (karmphal), without any knowledge. The work may be good or bad and so its result. The living entity is ignorant about all this. The victimization of action and reaction can be obviated by practice of yog. Lord Krishna advises to practice yog:

Yogasth kuru karmani sangamtyuktva dhananjaya I
sidhyasidhyoh samo bhutva samatvam yog uchchyate II 48 II

(Gita 2).

Buddhiyukto jahatiha ubhe sukrutdushkrute I
tasmad yogaya yujyasva yogah karmasukoushalam II 50 II

(Gita 2).

Duty be performed with un bias mind. Action performed with no expectation of its result. It is yog attitude. Actions performed with devotional service to super soul the result of actions good or bad does not affect the person. Yogi is considered well established when he is devoid of all material desires, with disciplined mental activities due to practice of YOG. He is like a flame of a lamp in a windless place. He is steady in his transcendental meditation (Gita stanza 18 and 19).

What science is involved in this statement of Gita?

Yogi's mind-set (CHITTA) is calm and still due to practice of yog and transcendental meditation. The practice is repeated cultivations on the mind. This cultivation has effect on the behaviour of yogi ('as stated earlier in this chapter"), which changes material body form of the yogi. The repeated cultivations change the arrangement of quality particles of DNA in the cells so that more divine qualities can be altered for spiritual elevation amounting to KUNDALINI arousal (Gita 20). The stanza 21 of Gita explains how intellectual ultimate happiness is experienced by yogi.

Sukhamatyantikam yattbuddhigrahyamatindriyam I
vetti yatra na cheivayam sthitshchlati tatvatah II 21 II

(Gita 6, 21).

The extreme and ultimate divine happiness (SUKHAM ANTYANTIKAM) at intellectual level beyond all senses (BUDHIGRAHYAM ATINDRIYAM). To understand this one must understand the origin of the word SUKH – the happiness. SU means good and KHA means Aakash element (like the word KHAG mean bird or one who goes to Aakash). What relation is of happiness to Aakash? All other elements like earth, water, air, tej are in Aakash tatva and vibrate in aakash tatva. Space or Aakash is in all these elements. Happy vibrations obtained on earth-PRITHVI TATVA, have relation with AAKASH TATVA. Therefore, happiness is SUKHAM. Vibrations in Aakash principle. Scientifically radiations on earth are stable due to sun radiation. The cultivations of yogi on earth also rise to Aakash tatva. Hence it is extreme and ultimate happiness[12].

As expressed earlier that PRAN energy flows through channels called IDA, PINGLA and SUSHUMNA. As per modern science these channels are represented by Right and Left SYMPATHETIC nerves on both sides of spine in retroperitoneal space. SUSHUMNA is represented as SPINAL CORD starting from BRAIN stem Medula Oblogata (part of brain) to sacral region. To stimulate these nerves or NADI procedure has been described. Before that let us understand the Right and left sympathetic chain – IDA, PINGLA nadi. This is as per modern science of Anatomy.

IDA nadi is left sympathetic nerve on left side of spine in retroperitoneal region. It is also called "CHANDRA nadi". Healthy person breathes alternately through left and right nostril or through both nostrils. When breathing is predominantly through left nostril, it is called "CHANDRA SWAR". PINGLA NADI is Right Sympathetic nerve on right side of spine retroperitoneal space (as per modern anatomy). When breathing is through right nostril it is called "SURYA SWAR". SUHUMNA nadi is with in the central space of spinal column. When breathing is through both nostril it is called "SUSHUMNA SWAR". It is very systematic arrangement of natural breathing with automation. Which nostril breathing is prevalent at what time and when, has also been calculated in Yog shastra (Details may be understood from original works). CHANDRA SWAR is mainly TAMAH dominated (Tamah pradhan). It is cool natured, like night atmosphere, during rising moon phase of 15 days. SURYA SWAR is RAJAH dominated (rajah pradhan). It is hot natured, like day atmosphere, during declining moon phase for 15 days. SUSHUMNA SWAR is SATVA dominated (Satva pradhan). Breathing through both nostrils. It is morning and evening time. *(PANTAJAL Yog Pradeep, SamadhiPad, SUTRA 34, P. 225-235. Code 47. Gita Press Gorakhpur. ISBN 81-293-0011-7).*

Control of PRAN and APAN vayu is mentioned in GITA Chapter 4 stanza 29:

Apane juhyati pranam prane pranam tathapare I
pranapangati ruddhva pranayamparayanah I
apare niyataharah pranan praneshu juhyati II 29 II

(Gita, 4).

With breath control and meditation yogi remain in trance. **Control of diet has also been stressed.** By control of PRAN process, yogi can control senses and mind set (CHITTA). That is why the PRANAYAM is essential part of YOG SADHNA or yog practice (yog means union with supreme). How deep and scientific study of Sages and Rishis of vedik period can be understood by the following facts:

Diva na pujayelingam ratravapi na payjayet I
sarvada pujayelingam divaratranirodhatah

Quoted from (Pavanvijay Swaroday). Patanjal Yog Pradeep, Code 47, Gita press Gorakhpur. P. 231.

Yog practice be carried out in the evening and early morning. Because during day when SURYA SWAR is dominant which is RAJO GUN dominant and during night when Chandra swar is dominant it is TAMAH GUN Dominent as stated above. During Joint period of day and night in evening and in early morning during joint period of night and day is proper for practice of yog and pranayama. During this time-period the breathing is through both the nostrils called as SUSHUMNA SWAR when the Satwik is predominet. Sutra in Sanskrit is stated above

Sushumna Nadi is superior to Ida and Pingla nadis. It is Spinal cord in modern science and lies in spinal column from head and neck to Sacrum. The **MULADHAR CHAKRA** which is Pelvic Plexus is situated two fingers above anal region and two fingers below root of Penis. Below this there is a micro (SUKSHMA) Traingular mechanism called YONI MANDAL. From left angle of triangle IDA NADI, from right angle PINGLA NADI and fro top angle of triangle SUSHUMNA NADI comes out. This is called "separate Tripartrite" NADI (MUKTA TRIVENI). Because the three nadies are separate from each other.

These nadis are given names of three rivers as GANGA, YAMUNA and SARASWATI respectively. These at Medulary Plexus (AGYA CHAKRA), unite and called YUKTA TRIVENI, union of three rivers (TRIVENI SANGAM). This is at par with modern science as well. In Medula Oblongata the fibres cross from left to right and from right to left called decussation of Pyramids (explained below).

Ida Bhagirathi ganga pingala Yamuna nadi I
tayormadhyagata nadi sushumnakhya saraswati I
trivenisangamo yatra teerthrajah sa uchchyate I
tatra snanam prakurvit sarva papeih pramuchchate

(PANTANKAL YOG PRADEEP. Gita press Gorakhpur, code 47. P 241).

In Modern science study of brain stem the Medulla Oblongata which is lowest part of brain stem and continues as spinal cord below. **Anterior part of medulla displays as longitudinal elevations on either side of midline. These are PYRAMIDS. These pyramids are made up of CORTICO SPINAL fibres. These fibres cross the midline i.e. Left side fibres cross to right side, and right to left side. This crossing is known as Pyramidal decussation (see figure).** This is Yukta triveni at AGYA CHAKRA in Vedik Literature as mentioned above (this coincides with and in line with place between two eye brows or Bhrkuti madhya).

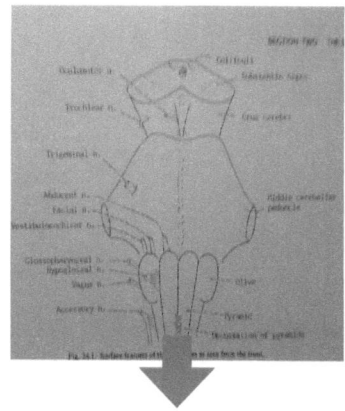

Crisscross lines are crossing of fibres.

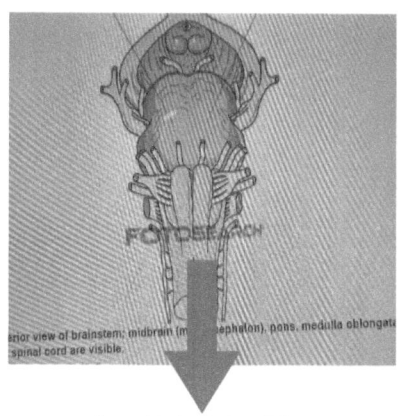

pyramids of Medulaoblongata on both sides of midline.

With practice of yog and pranayama "mind – set" can be regulated and controlled with focus on above plexus – CHAKRA. Hence must be practiced (Gita 4, 29 – quoted above). Question is how thought focused on AGYA CHAKRA {or different CHAKRAS} can control mind-set. One knows that the mind has wonderful control on body senses. As discussed previously that eyes do not see when mind is not focused on object of vision. Ears do not listen when mind is not focused on listening. When a lecture is delivered in a classroom, and student's mind is wandering elsewhere, the student is unable to understand the subject of lecture. Similarly, it is practically proved that the CHAKRA (or nerve plexuses), get stimulated when mind is focused on that particular Chakra, during meditation. Hence Gita advocates the fact in Chapter 4, stanza 29. Anatomical details of Sushumna nadi – spinal cord, described are at par with modern anatomy of spinal cord (see figure-shows Brain and spinal cord in spinal column).

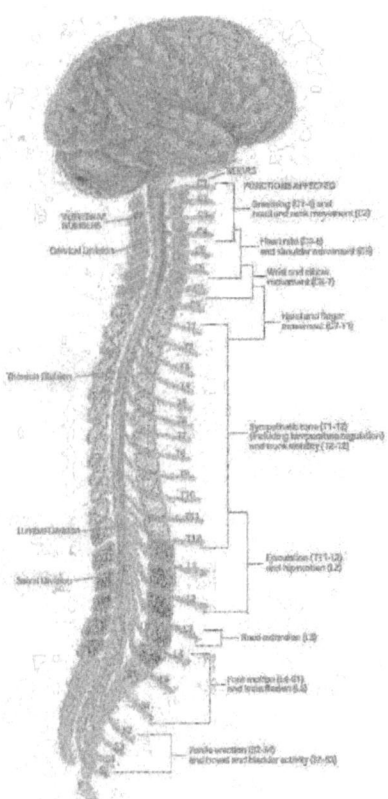

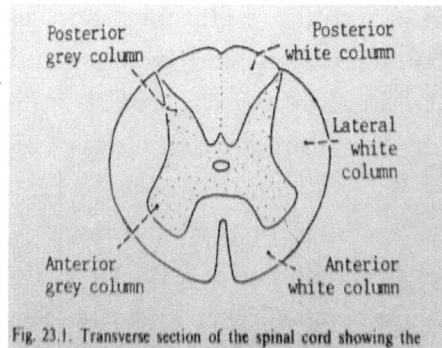

Fig. 23.1. Transverse section of the spinal cord showing the arrangement of the white and grey matter.

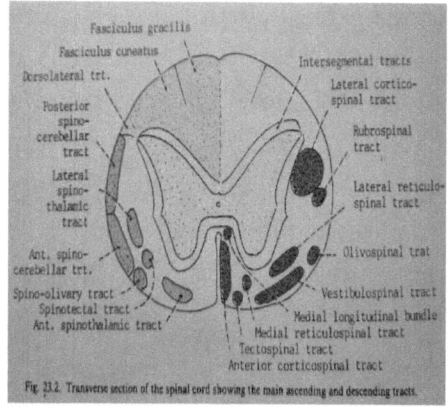

Fig. 23.2. Transverse section of the spinal cord showing the main ascending and descending tracts.

Transeverse section of spinal cord showing grey and white matter and central canal. Showing ascending and descending tracts in spinal cord (Modern science).

Inside of Sushumna Nadi: The meaning of Sanskrit word NADI is flow channel. The life force flows through the Nadi. Sushumna Nadi permits flow to and from brain to all body. It has inside spinal cord a nadi named "VAJRA NADI" inside the SUSHUMNA nadi. Inside Vajra nadi there is another nadi called "CHITRANI NADI" Inside Chitrani Nadi in the centre there is "BRAHM NADI". TheseNadis are enlightened and have hyperphysical energy. These are the seats of micro body (SUKSHMA SHARIR) and micro pran 'SUKSHMA PRAN' as stated in earlier chapters. There are many energy centers on Suahumna Nadi. Many micro channels unite in these centres. Out of these many centres the main centres are Seven. The names of these centres or chakra are:

MULADHAR, SWADHISHTHAN, MANIPUR, ANAHAT, VISHUDHI, AGYA and SAHASTRAR.

These are the abodes or districts of seven abodes (Saat lok) described earlier. These are included or contained with different enlightenments and electromagnetic forces. In normal conditions this is like a lotus flower opened and petals droped downwards. With meditation and DHYAN get stimulation and blossom up with petals upwards and flower gets openedup. Then only these centres are illumined and get energized. Every CHAKRA has different amazing and hyperphysical energies. Following description with English words used is only symbolic, approximate and presumptive. English names of plexuses are only indicative and gestural.

Like knowledge of "mind-set" it is essential for yog practitioner to know about the "PRAN". It must be clear here that the PRAN is not the breath only, as considered by common people. It is neither the "Soul Element" as considered by modern scientists. **But the PRAN is that element by which the breathing action and all other body actions, in a living body take place.** In the beginning of Universe in nature the five basic elments (Panchbhut – earth, water, fire, air and aakash) and everything in nature are manifested due to interaction of Aakash and Pran Shakti. The life of living beings is due to this Pran Shakti. At the time of annihilation all are devoid of this pran Shakti support and get unmanifest into their cause – the Aakash. This is described in Chhandyog upnishad 1. 9. 1. Manifestation of material nature (Bhutotpatti) and its dissolution in Aakash and Pran:

Asya lokasya ka gatirityakash iti hovach sarvani ha va Imani
bhutantakashdev samutpdayante aakasham pratyastam
yantyakasho hyeveibhyo jyayanakashah parayanam IIIII
Chhandyogopnishad Chapter 1, Khand 9.

(Chhandyogopnishad code 582, Gita Press Gorakhpur. Chapter 1, Khand 9, stanza 1. Page 91).

CHAPTER SEVEN

SCIENCE OF DNYAN AND VIDNYAN YOG
SCIENCE OF PRANAYAM AND YOG
(SUBSTANCE OF CONVICTION)

Abstract: According to rishi VEDVYAS author of Gita, science of SANKHYA YOG, is superior most in all knowledges. As per the Sanskrit literature the universe is created by Brahm. The science of SANKHYA YOG DNYAN, or VIDNYAN of Sankhya Science is greatest in all branches of knowledges. The nature is of 8 varieties – ASHTDHA PRAKRITI. (5 ELEMENTS, 3 FACULTIES OF MIND – buddhi, ahankaar, chitta. – Gita 7. 4). With the help of these Brahm created JEEVATMA. This existence is illusion of MAYA. This lures all living beings of the universe. It entangles JEEVATMA. The existence is illusion. This illusion can be overcome by proper karm or deed as per the body provided by the nature. This takes the jeevatma to the goal of emancipation by becoming himself BRAHM.

The opening statement by Shri Krishn in 7th chapter is "with devoted mind to super soul (me i.e. Lord Krishn) practice of Yog, one can understand Super soul doubtlessly".

Mayyasaktamanah partha yogam yunjnmadashraya I
asanshyamsamagram ma yatha dnyasyasi tachchhrunu II 1 II

(Gita 7, 1).

Surrendering mind to super soul while practicing Yog advised by Shri Krishna, to Arjun. Without any doubt one understands the supreme soul, and one becomes knowledgeable, is a fact listen from Me.. Why the stress is on Attachment of Mind to Super soul? The knowledge of "Absolute Truth" completely is possible only with such 'wholehearted' attachment. Otherwise the knowledge of soul is only partial. A small fraction of absolute truth or Supreme soul is in every living being. What is MIND reflected in stanza 1 above? There is massive literature available on mind, in Sanskrit literature and in Gita. In YOGAVASISTHA RAMAYANA, Sage VASISTHA tells RAM as student about mind.

Jayate mana eva vivardhate I
samyagdarshandrushtya tu mana evahi muchyate II 14 II

(Yogvasistha Ramayana 5, 11. 14).

Mind takes birth, it grows. Existence of mind is in fact imaginary. It thrives because of consciousness. Oneness with pure consciousness becomes eternal brilliance.

Mana evamsatkalpam chitprasaden jivati I
bhavyan vishravmeveikam chintametya chidapyut II 49 II

(Y.R. 13, 49).

Mind is not substantial but imaginary. Mind is not a distinct entity. Apparent sequence of thought is considered as mind. When imagination (SANKALP) of mind decides that "I am pure consciousness", it turns pure and becomes pure consciousness.

Chittaswabhavparamushta spandshaktirsanmayi I
Kalpana chttamityuktya kathyate shastradrushtibhi II 51 II

(Y. R. 13, 51)

Because of the transparency of mind, it is 'non-existent' and hence it brings about an illusion. Such duality always surrounds it.

In Gita stanzas 4, 5 and 6, of this chapter Lord Krishna reiterates that Super soul is represented by two types of energies called APARA and PARA:

Bhumiraponalo vayuh kham mano budhirev cha I

ahankara itiyam me bhinna prakritishtadha II 4 II

(Gita 7. 4).

These 8 types of energies are Apara energies – (Earth, water, Fire, Air, Space, Mind, Intellect, and Ego).

Apareymitstvanya prakriti vidhi me param I

jeevbhutam mahabaho yayedam dharyate jagat II 5 II

(Gita 7. 5).

Apart from Apara energy, which is inferior to superior energy called Para energy. The Nature (Prakriti) is contained, controlled, and manifested by PARA energy, which is enjoying Apara facility [or Para is enliving (– idam dharyate), physically manifest world which is Apara energy].

In short, Physical (manifest) world is due to APARA SHAKTI or Physical energy and enlivened and activated by PARA SHAKTI or Spiritual or soul energy. Therefore, concentration on super soul knowledge gets revealed to practicing person. All innovative scientists have experienced this since time and again.

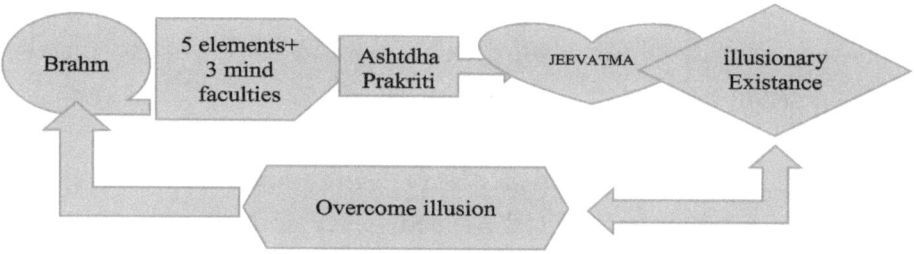

Brahma Cycle: creation of Jeevatma from Brahm and becoming brahm with efforts.

Practice of yog is advocated by Lord Krishn to understand and acquire knowledge about super soul in first stanza of Gita, chapter7:

Mayyasaktamanah parth yogam yunjnmadashrayah III II

Similar expression is in Mundkopnishad (1, 4.) on PARA and APARA knowledge.

Tasmei sa hovach I

dvei vidye veditavye iti ha sma I

yad brahmavido vadanti para cheivapara cha II 4 II

(Mondkopnishad. I, 4).

PARA is ultimate knowledge, which enlightens one about Supreme soul. APARA is limited knowledge about nature and surrounding manifest material world. Ultimate eternal supreme soul is revealed by spiritual knowledge (para vidya).

In next few chapters the science of YOG and how to achieve the knowledge successfully has been described. Procedures and techniques are described in the form of examples. Science of YOG and PRANAYAM is described elsewhere in Sanskrit literature. Therefore, to understand these examples AUTHOR has referred to original works elsewhere. PATANJALI who is the greatest author on YOG SHASTRA OR YOGALOGY. PRANAYAM AND YOG ASANS are hand and glove to each other. Both are complementary to each other. PRANAYAM is PRAN + AAYAM i.e. control or regulation of taking in air and breathing out air – to and fro-from lungs through respiratory tract and nostrils. In other words, PRANAYAM is breathing exercise with a technique.

It needs a clarification on "breathing techniques" PRANAYAM and some sitting postures referred as ASAN. At the same time Pran which is confused with air exaled and inhaled. Since control of breathing is PRANAYAM and it deals with breathing techniques governing air in and out of lungs, therefore it is considered air containing Oxygen and Carbon di oxide as per understanding of modern science. In fact, with inhaled air, one breath in or one inhales, more vital things needful for sustaining life and life's processes. It is known that surrounding Earth there is space (– AAKASH – in Sanskrit). It is not empty. It has AETHER which contains much more than simply air and other ingredients, still unknown to modern science. It is a subject of further research. It will suffice here, as per VEDIK knowledge, to say that it has an energy which sustains life. Best example for this is that it is known that watering plant with water from river, well, Lakes or tap water sustains life in plants but rainwater, with thunderous sound of clouds during rain rejuvenate plants with new life. With "big bang" of clouds or "OHM NAAD" life energy dissolves in rainwater drops which energizes the soil on earth and in turn the plants. **This is life force or PRANIK SHAKTI referred as PRAN WHICH IS INHALED ALONG WITH AIR. In earlier chapter ten types of PRAN and their nomenclature have been described (Chapter 4). PRAN is life energy. Different PRANIK SHAKTI regulates different functions of the body.** These functions are described earlier. PRANAYAM techniques in collaboration with sitting postures – ASAN help user to accomplish rejuvenation and transcendental liberation. One knows that if man or animal strangled for more than three minutes continuously the death is inevitable but the PRAN is present in tissues and organs after death. This PRANIK SHAKTI called as DHANANJAY which is responsible for maintaining nourishment to tissues and organs after death. That maintains the vitality and nourishment of tissues and organs. **That is why transplantation of tissues and organs is possible even after death. Not only that the animal consumed by predator receives the needful nourishment. It is fractionally BECAUSE OF PROTEIN SOURCE FROM MEAT but mainly due to the 'pranik Shakti' named as DHANANJAY PRAN. This fact was known to saints, sages, Rishi and Muni's of Sanskrit speaking civilization.** Accordingly, the various types of PRAN have been classified and named as per their function and utility. This has been described earlier. To author's understanding and knowledge, this is still a subject of research for the modern science and researchers, to unravel what exactly PRAN is? Space (or AKASH in Sanskrit) contains ether, as stated earlier, has not only gasses but has vital force essential for life in living beings. What this vital force is still unknown. It is much more complicated than simply Oxygen, Hydrogen, Carbon di Oxide, Nitrogen, Ozone, Helium etc. known to modern science. Vedik literature goes further to say that the individual soul factor getting attached to raindrops, falls on earth (stanza given **below**). It mixes with soil and remains there for some time {the duration is dependent upon the fruits of work or actions performed (KARM PHAL) of the soul in previous incarnation} and along with grass and vegetation eaten by herbivore animals and man. After entering into the blood circulation, it gets to reproductive organs, and through this after union with female, new life is created. See stanzas below from Madbhagwat-Shaktipeethank:

Swakarmashato jeevo niharkalaya yutah I
patitva dharaniprushthe vrihimdhyagato bhavet I110II

Sthitva tatra chirambhuktva bhujyate purusheistatah I
Tatah pravishtam tadguhyam punso dehe prajayate II 11 II

Retastena sa jeevopi bhavedretogatastada I
Tatastriyabhiyogen ritukale mahamate II 12 II

(Devi Puran part of Mahabhagwat – Shaktipeethank, Gita Press Gorakhpur year 79, no. 1. Publication year-2005) Chapter 17, PAGE 157-158.

Depending upon the work-action effect of the living being, uniting with dew drops (Niharkalay yutah), falling on earth (patitva dharana prushthe) and reaches plants and vegetation. It stays there (sthitva tatra chiram) and enjoys or suffers (bhuktva bhujyate) for various time-span (period), proportionate to the KARMPHAL – results of deeds. It is consumed by other living beings. It then enters the reproductive sense organs (Tatah pravishtam tadguhyam). **After union with female** (Tatastriyabhiyogen) **a new life begins** (Retastena sa jeevopi bhaved).

A comparison between BHAKTI YOG and renunciation is described. Mental speculation and renunciation of work is not that important as the devotional action. Surrendering whole heartedly to supreme soul, one attains peace from the pangs of material miseries. Performing duty without attachment and surrendering results of duty/work to supreme soul, one is free from bondage and unaffected by sinful actions. It is like lotus leaf existing in water but unaffected and un-adhered by water. To achieve this goal PATANJALI in his book on YOG, HAS DESCRIBED EIGHTFOLD PROCEDURES, the same have been mentioned in short so that meaning of GITA stanza can be scientifically understood better. These are named as ASHTANG YOG. These eight types of principles of 'ashtang-yog' are:

YAMA, NIYAM, ASAN, PRANAYAM, PRATYAHAR, DHARNA, DHYAN, SAMADHI.

In seventh chapter of GITA it is stated that one in many 1000 only one can reach to perfection. Those who have achieved perfection, hardly only one really knows about supreme soul, in true sense. Generally, majority of people are interested in sense enjoyment of material world like animals. This does not provide true happiness but engage the person into camouflaged world i.e. MAYA in Sanskrit. The transcendental knowledge, understanding self, Super self and the process of understanding DNYAN YOG, union with knowledge, DHYAN YOG – meditation. Supreme soul is cause of all causes. Material energies of nature like earth, water, air, fire, mind and ego, ether, intelligence are energies which are separated from super soul. Apart from this a superior form of energy is in all living forms in universe (universe in total is "living", in fact all material world is living. It has been explained previously). Super soul is both, original seed of all existence as well as dissolution of this universe. In material world the modes can be goodness, Passion, or ignorance, (Satva, Rajah, tamah) but behind all this is supreme soul – the absolute truth. Its nature is imperishable, infallible, unborn, and supreme.

In SANKHYA terminology, review the meaning in Gita stanzas 4, 5, 6, of chapter 7, in context with RIGVED-NASDIYA sukta. Principle elements in cosmos and its evolution as described in RIGVED is below:

RIGVED NASDIYA SUKTA: **Deals with cosmic evolution.**

Seven mantra from Nasadiya sukta:

Om nasadasinno sadasittadani nasidrajo no vyouma paro yat I
Kimavrivah kuha kasya sharmannabhah kimasidgahanam
gambhiram II 1 II

There was nothing before the manifestation of universe. The beginning was from zero. No space, wind, but only darkness existed. Why, that plasma, all pervading, deep and profound?

Na mrityurasidmrutam na tarhi na ratrya ahna aasitpraketah I
Aanidavatam swadhaya tadekam tasmadhanyanna
parah kinchnas II 2 II

There was no death or immortality. No earth no life. No immortals like in heaven. No day or night. But for that breathless one breathing on its own. Only eternal existed.

Tam aasittamasa guhalmagre praketam salilam sarvaidam I
Tuchchhyenabhvpihitam yadasittapasastanmahinajayateikam II 3 II

In the beginning only darkness was there. Eternal element – (PADARTH-dravya) existed without any form.

Kamastdagre samavarttadhi manaso retah prathamam yadasit I
Sato bandhumasati niravindanhadi pratishya
kavayo Manisha II 4 II

Creator of universe thought of establishing universe, from unmanifest to manifest world. In the cosmic mind, all pervading desire, the primal seed made its first appearance and the wise men, seeking deep in their heart Could see the link between 'that is' and 'that is not'.

Tirashchinno vitato rashmireshamadhah swidasit I
Retodha aasanmahiman aasantsvadha avstatprayatihi
parastat II5II

With such thought rays emerged and energy established, due to which using the nature/element creation started. Reins of the link, a grid of crisscross lines, Holds all the seeds and mighty forces, Microcosmic forces within and macro forces out above.

Ko adha ved ka iha pra vochatkut aajata kut iyam visrushtihi I
Arvagdeva asya visarjnenatha ko ved yat aavabhuva II 6 II

No one can say when and how the world manifested. Because the scientists and knowledge-able came into existence after that. Before the manifest universe what it was and what was there and what is the cause of creation, is an all enigmatic.

Iyam visrushtiryat aababhuva yadi va dadhe yadi va na I
Yo asyadhyakshah parame vyoumantso anga ved yadi
va nav ed II7II

What is the source of creation, who is responsible for creation, Who is director, observer, surveyor of this creation established in space or (in heaven). Oh! scientists of the universe find out, who else can find out other than you!!! That one, out of which the creation came May hold the reins or not, Perceiving, all from above, 'That' one alone Knows the beginning – may not know too.

On analysing one by one as per SANKHYA YOG and RIGVED-NASDIYA SUKTA:

In SANKHYA Shastra and YOG Shastra two eternal elements are assumed:

1. Spiritual element (jeevatma) and
2. Material element (prakriti).

The spirit which is inactive, knowledgeable, non-smearable, and non-attachable, while the other is material element. Both are eternal elements. Spirit is eternal energy while material world is (trident root elements – SATVA, RAJAS, TAMAS) eternally effective and active. In GITA it is emphasized in following stanzas:

> *Bhumiraponalo vayuh kham mano budhirev cha I*
> *Ahankara itiyam me bhinna prakritirshtdha II 4 II*
>
> **(Gita 7, 4).**

> *Apareymitstvanya prkritim viddhi me param I*
> *Jeevbhutam mahabaho yayedam dharyate jagat II 5 II*
>
> **(GITA Ch 7, 5).**

Earth, Water, Air, Fire, Ether and Mind, intellect, Ego (false) are eight types of separated energies of super energy. These are inferior energies. There is another energy – Superior energy – which in spirit form present in living entities. This superior energy exploits and uses the inferior material energy for various purposes. All created beings covered under these two basic elements. Both have origin and dissolution effected by super soul/energy. In reiteration to this the 10th stanza says:

> *Beejam mam sarvabhutanam viddhi parth sanatana I*
> *Budhibudhimatamasmi tejasvinamham II 10 II*
>
> **(Gita 7. 10).**

Super soul (Shri Krishn used the word "I" highlighted and underlined words in above stanza) is seed of origin and existence, intellect in intelligence, power in powerful in this universe and light in TEJ.

Supreme soul is PARABRAHM (absolute truth), FRACTION OF THIS IS BRAHM which is in the form of seed is present in moving and inert (living and non-living word is not used here because every material entity is living). The original root is PARABRAHM. The same version is in KATHOPNISHAD.

> *Nityonitynam chetanshcha chetnanam eko bahunam yo*
> *viddhati Kaman I*

tamatmstham yenupashyanti dhirakhstesham shanti shashvati
netaresham II 13 II
(Kathopnishad Chapter 2, Valli 2, stanza 13).

As per stanza 4, the eight elements, (Panch Mahabhut (bhumiraponalonilonabhah) and mind, intellect, ego (Mann, Buddhi, Ahankar) total 8) are separated energies of super soul as stated above. What SANKHYA and RIGVED means by these terms. NASDIY cosmic evolution:

AAKASH: Aa=together and KAASH=ENVELOP. It is above space and time. It is first expression of super soul. It is evolved first before any other element or BHUT (manifest, unmanifest). It can be perceived by sound vibrations and bluish tinge colour. It is beyond electronic stage-TACHYON which has speed limit. 'Aakash-Tatva' above speed, space and time. Hence above Tachyon. Speed of Tachyon is 5 times the speed of light.

(European Physical journal C 76, by article, "Birth and death of Universe. "number 709, 2016. DOI https://doi.org/10.1140/epjc/s10052-016-4577-8 authors: Fried, HM., Gabellini,)

VAYU: After Aakash expression of super soul is VAYU (not air or gasses). An expression that can travel all over within Aakash – i.e. space. Scientists believe that there are numerous waves beyond the orbit of the electrons, pulsating to eternity (see reference below). Vayu is perceived by touch and smoke like tinge of colour. Speed and existence of VAYU is beyond imagination hence it is called as VAYU. It is second in succession of cosmic expression.

(Journal of Geophysical Research: Space Physics. Volume 115, Issue A10. Magnetospheric Physics. Free Access. "THEMIS observations of electron cyclotron harmonic emissions, ULF waves, and pulsating auroras". Jun Liang, V. Uritsky, E. Donovan, B. Ni, E. Spanswick, T. Trondsen, J. Bonnell, A. Roux, U. Auster, D. Larson. First published: 19 October 2010. https://doi.org/10.1029/2009JA015148).

In author's opinion, existence of Vayu (wind, air) on earth and around the planet earth could be as follows:

When earth separated from its origin and was manifested, as hot ball of gasses. With rotation around its own axis the gravitational force the gasses kept clinging to the planet. On cooling and solidification of gaseous matter, solid part of earth came into being. But some gasses still, remained circulating around the planet. Movement of gasses likely to be due to rotation of earth around its axis, unequal heat on different solidified matters on earth, giving impetus to the flow of wind particles. Vegetation on earth added some new gasses and altered the gaseous composition of gas atmosphere around earth. This is an ongoing process of nature (prakriti). All is under the effect of time scale or time cycle (kaal chakra). Wind movement is also influenced by movements of other stars of the galaxy.

TEJUS: Tejus is speed. Speed of light is 300,000 Km/second. Sankhya philosophy believes that tejus has quality to create. It can be perceived as like fire flame. Modern science accepts that un-restricted or un-bonded energy is **light**. Restricted or bonded or bottled energy is **electron (within orbit).** It is third in succession in cosmic evolution event. (NASDIYA SUKT 3 stated above).

AAP: Aap is obtainable material therefore called as-aap. It is obtained from bonded energy – light energy or electron which move with great speed. Matter is obtained from the electronic stage of cosmic expression. The aap is the 4th in succession from Aakash tatva (4th expression of Nasdiya sukta as above). Aakash tatva create liquid state as the expressed elements. With special affinity in electrons or elements, the liquid is called as RAS (liquid state). Taste, which is a quality of aap tatva. It is perceived by whitish complexion. It radiates light rays. It is the primal seed of cosmic expression.

PRITHVI: Analytically meaning of Prithvi is PRITH(AK) =separate and VI = give birth. With Aakash AAP gives birth to other gross stage of expression called PRITHVI. It is 5th expression of cosmic evolution. It is impregnated with innumerable materials, elements, compounds and living beings. Because of gross visibility it is called as PRITHVI. It is recognized due to its GANDHA (not smell as popularly understood). In SANKHYA philosophy it means = goes to incorporate. To elaborate the meaning of GANDHA, the earth due to its special property of incorporation the seeds of different plants perceive different quality. In a piece of land two different seed with other exposures being same (atmosphere, light, air, sunlight, water etc. being commonly the same) exhibit different taste and colour. Sugar cane = sweet, Bitter gourd = bitter, Mango = sweet and sour, Lemon = sour.

Seventh stanza of the Nasdiya sukta is highly scientific. After explaining the creation of cosmos.

Rigved states in sukta 7.

That one, out of which the creation came May hold the reins or not, Perceiving, all from above, 'That' one alone Knows the beginning – may not know too".

These 5 elements in SANKHYA PHILOSOPHY (PANCH MAHABHUT) have unique nomenclature. In English there are no proper synonyms known to author, hence used as it is.

PARAMBRAHM IS THE ETERNAL AMONG ETERNALS and supreme LIVING ENTITIES. Living knowledgeable, among knowledgeable i.e. – AATMA. This probably, solves the perpetual question that what first existed egg or hen. Thus, the physical energy is controlled and directed to work according to the wishes of the absolute truth. The modern science is striving to unravel this fact by scientific proofs. In 15th stanza (quoted below) GITA explains that those miscreants who are lowest of human race are grouped into 4 different types.

1. Foolish – MOODH,
2. Lowest of mankind-NARADHAM,
3. Mankind whose knowledge is stolen by illusion, MAYA-APARHITA-DNYAN (see below-Maya + apahrit + dnyan),
4. Demonic-ASUR.

Na mam dushkritino mudhah prpadyante naradhama I
myyapahrutdnyana aasur bhavmashritah II 15 II

(GITA, 7, 15).

All these members of mankind are grossly involved in work for material gains and fruits of work, are unable to reach the transcendental level and fail to rise to higher levels. After many birth and deaths true knowledge comes to them. The liberation or knowledge is not denied to these elements of human race or even animals. *This clears the view that the repeated birth and death is for the welfare of individual. Repeated births are for liberation of the living beings.* The enlightened souls are rare, hardly one in many millions (Gita ch. 7 stanza 3). Nature has provided chance for improvement till annihilation.

In next stanza the pious humans are described. These pious humans practice transcendental meditation are of 4 types: Those who are

1. Distressed,
2. In need of wealth,
3. In search or inquisitive of knowledge or
4. In search of absolute truth.

The first three are having selfish motives. The last one is of pure devotional aptitude and is the best type and dear to supreme soul. Such persons are sure to get liberated.

Such scientific and analytical statements are to de-stress humans and persuading all for pious living for betterment of social living and societies – human race at large. Science of society maintains that Human society consists of different people. As stated above, humans can be grouped into as per their – Genetic and intellectual development, which decides their groups. Lowest, Middle order and High order category as per mental intelligence. Basically, main two groups can be mentioned in a broad manner – One group indulged into material gains, greedy addicted to sense gratification. The other is with high degree of intellectual ascetic and free from material gains or greed, non-selfish, merciful and are evolved. Selfishness, greed with sense gratification for self, make people ruthless and to achieve and satisfy ego, this makes them merciless and even cruel. Without caring for others in the society, such people take the help of politics and power. Through such attempts the ego is satisfied. On the contrary pious persons act and behave without ego and pride. As per GITA the mind-set is totally regulated and controlled by the super soul as stated in above chapters. The diagram shown in chapter 8, page 208, The Holy Science, by Shri Yukteshwar Giri, has Constellations and Galaxies which comes in rotation and repeatedly. Even creation and annihilation of BRAHM is also cyclical.

Does that mean nothing can be done to the mind-set of humans of the society?

Gita very clearly answers this question. All category of persons can practice and achieve piousness and transcendental meditation, liberation, education and training. That is why in Sanskrit speaking civilization, 16 types of sacraments are advocated. (SOLAH SANSKAAR). Highest preaching of ISHAVASYOPNISHAD.

Isha vasyamidam >\ sarvam yakinch jagatyam jagat I
Tyen tyakten bhutrijthah ma grudhah kasyasvid dhanam II 1 II

Kurvanneveham karmani jijivishet shatam >\samah I
Evam tvyi nanya thetosti na karmam lipyate nare II 2 II

The whole universe is unsteady or moving, it is pervaded by soul fraction part of super soul, (everything is moving all living beings, plants (roots of plants move in search of water under ground and over the ground rotate or move towards sunlight), even a small atom having electron, proton, neutron are moving inside atom, stars, Nakshtra – group formations of stars, constellations, galaxies etc.). All material world is pervaded by super soul, along with super soul, keeping all this in mind, without greed, unattached to materialism of material world, use it unattached or without any kind of greed or attachment, material world belongs to nobody, therefore (Tyen tyakten bhutrijthah) god provided material world, enjoy and suffer both for carnal pleasure. (bhutrijthah =bhog=pleasure-grief). With such a mind-set live destined life of 100 years.

The scientific aspect of this has been explained earlier. The quality particle placement in chromosome coil can be changed with perseverance of good or bad cultivations which decides nature of the individual.

Instead of pious teaching some cults and faith teach cruelty to younger generation towards animals, plants or even incumbent humans of society. 16 sacraments start with conception – (GARBHADHAN) and end with cremation or funeral. For details reader is referred to appropriate literature as these are beyond purviews of this book. **In short the social science was developed on laws of the universe and nature of universe, psychological and scientific analytical basis. The social welfare was a big concerned and at the same time the progress of individual was also considered.** As per theory of DHARM, ARTH, KAAM, and MOKSH is concerned the highest aim of life after sense gratification in life, to achieve liberation or MOKSH, is also taken care of. Without renunciation and dedication of work to super soul and not caring for the work result etc. is only possible by adopting pious attitude, liberation is difficult and cannot be achieved, was well known to the rishis, sages and saints of that civilization. Even the prayer MANTRA dedicated to the formless, shapeless, pure NIRGUN, NIRAKAR, PURE super soul and requested to direct our intellect and mind-set in that direction. It deals with eternal hence SANATAN DHARM for example:

Om bhu rbhuvah swah tatsaviturvarenyam bhargo devasya dhimahi dhiyo yo nah prachodayat I
RIGVED (mandala 3.62.10).

Prayer to absolute truth (parabrhma), the creator of all universe, protector of universe, to stimulate intellect and bestow true knowledge upon us. There is not the least tinge of modern religion in any form in this prayer. The prayer is only for knowledge and awakening intellect. All this reiterates the eternal mindset of DHARM (not religion).

Pious preaching is the duty of elders towards the younger generation for pious human society. True knowledge physical or spiritual is possible with peaceful, pious and tranquility in human society. With peace and tranquility in human society overall progress is inevitable. Scientific basis has already been explained. This is probably the reason for overall progress and prosperity of Sanskrit speaking civilization of Vedik period.

CHAPTER EIGHT

ETERNAL BRAHM YOG
SPACE SCIENCE DISCOVERY OF ZERO
UNDERSTANDING UNPERISHABLE TRUTH
JYOTISH SHASTRA

Abstract: This chapter basically is on science of body and science of how to behave as per the 'dharm' the nature has provided. This is to get perfect knowledge about the self and eternal truth. Yog of unperishable 'brahm' or Akshar Brahm, reveals scientific information of high calibre. Scientific explanations about experiences after death have been scientifically explained, in the form of, dialogue.

Arjun wanted explanation about what is Brahm, Self or Aatm, Material world and Divine supernatural agencies from supreme authority that is Lord Krishn (he addressed him as PURUSHOTTAM). He sought clarification about, at the time of death when all senses become non-functional how one can concentrate on supreme soul (Gita 8. 1 and 2).

Kim tad brahm kimdhyatmam kim karm prushottam I
adhibhutam cha kim proktadhideivam kimuchyate III II

(Gita 8, 1).

Adhiyadnyah katham kotra dehe sminmdhusudana I
prayankale cha katham dnyeyosi niyatatmabhihi II 2 II

(Gita 8, 2)

Three explanations demanded in stanza 1, by Arjun include:

What is BRAHM and AADHYATM?
What is AADHIBHUT or material manifestation?
What is ADHIDAIVAM or eternal knowledgeable life?

In 2nd stanza the explanation about science of ADHIYADNYA (or action with purpose and not sacrifice as mentioned earlier). What is the knowledge of existence after death?

Gita is written by Sage VEDVYAS. The creative intelligence of writer can be understood by the questions asked by Arjun and answers by Shri Krishn. Highly illumined and knowledgeable writer creates question queried through Arjun, which is relevant even today. The answer generated to satisfy not only Arjun but all humans, spoken by shri Krishn. Let us discuss question-answer one by one.

Lord Krishn answers in stanza 3 of chapter 8. The Brahm is that which never perishes, or which is eternal. The entrapped soul in material body (or-Jeevatma), is called as Adhyatma (Brahm). Original Brahm has tendency to expand and multiply through evolution process. Eventually enters many species – Virus, Bacteria, insects, other living entities. The capacity to trace origin is in humans due to developed brain. Capacity to trace origin is Adhyatm. Obviously, it is evolutionary concept of humans. Or it is nature-SWABHAV of humans. SWA means own and BHAV means concept. This Nature process of knowing original stage is Adhyatm, eventually the creative part of nature is KARM. Action pertaining to development of living material bodies is Karma (fruitive activities). This is scientific explanation of first line of stanza 1, stated above, the question asked by Arjun.

The materials created and destroyed and under control of nature are ADHIBHUT. Prakriti or nature in the form of life or living beings, which is enlightened is called ADHIDAIVA. Action which creates and protects Jeeva, is called ADHIYADNYA. The world is experienced by us by our mind, else there is no world. Though the mind is un-substantial, having no distinct entity, and is imaginary, but exists due to consciousness and has undoubtable presence, which can be experienced only. Power of mind controls the physical, material manifest body.

Chitchhakteh spandshakteshch sambandhah kalpyate manah I
mithyeiv tatsmutpannam mitthyadnyanam taduchyate II 88 II

Esha hyavidya kathita mayeisha sa nigadyate I
paramettaddnyanam sansaradivishpradam II 89 II
(Yogvashishth Ramayana 13. 88, 89).

Mind-set power and pulsating power, both working together effectively is mind. Even though it is unreal, hence the whole is considered as ignorance and illusion. It is AVIDYA or MAYA. But highly, powerful and infuses poison of creation and worldliness. Rishi Vasishtha further warns RAM that:

Anartamrutavsmin anaratsamudbhave I
sansar sambhrame yukta na tushtirna cha duhkhita II 35 II
(Yogvasishth 13, 35).

Illusion called SANSAR – world/worldliness, has incessant birth and death hence bemoaning is not called for.

Drishyamashrayasidam chettatsachittosi bandhvan I
drishyam santyajsidam chetttsachitto mokshvan II 1 II
(Yogvasishth 14, 1).

The visible world with concern in mind will lead one to bondage and if one ignores the visible SANSAR in mind, the same mind will lead to liberation.

Therefore, remain centred in sadhana in internal mind deeply, unconcerned in conscious mind. Free the mind from the influences of visible, is the state of liberation itself.

Mind is responsible for good or bad cultivations. It depends upon self-discrimination.

Cultivations imbibed during lifetime, become companion of the soul, which decide the life after death. It becomes the in-born culture for the soul. Therefore, Vedik literature emphasizes the cultivations, emotions, sentiments, tendencies etc. obtained during lifetime be such that helps upgradation for all time to come. (discussed elsewhere). Stanza 6 of chapter 8 of Gita, confirms this:

Yam yam vapi smaran bhavam tyaj tyante kalevaram I
tam tameveiti kounteya sada tadbhavbhavitah II6 II

(Gita 8, 6).

Same emotions and sentiments reoccur in mind at the time of death which have been cultivated during whole life.

LORD Krishn explains that in perishable material world, what is eternal is BRAHM. Transcendental eternal nature is self (or adhyatm). Manifestation of material nature, elements and living beings, is YADNYA KARM or productive work (karm) the super soul-BRAHM.

The Brahm manifests and has duration of life, it grows, multiplies, after the life span it unmanifest or dies. Therefore, Gita in stanza 8, 17. Scientific calculations about the time span or time cycle or Kaal Chakra. (citation in Sanskrit given below). It is known that the time is cyclical. On earth it depends upon the rotation of earth around star SUN. Similarly, time scale depends upon rotation of galaxy around centre of Brahm and so on.

Duration of Brahma's day and night are cited as example in this chapter. At the same time Lord KRISHN explains that, everything in this universe is temporary and perishable, but spiritual world is not perishable then why worry for perishables, on the contrary be a spiritual. Details are not cited in the stanzas, therefor for details one refers to main literature in BRAHM PURAN, MARKANDEYA PURAN, SHIV PURAN, JYOTISH TATVANK and MADBHAGWAT PURAN. For exploration of hidden science in stanzas of Gita, details from VED PURAN etc. are quoted.

Aabrahmabhuvanallokah punaravartinorjun I
mamupetyatukounteya punarjanma na vidyate II16 II

(Gita 8, 16).

All the stages of Brahm undergo repetition but after merging with supreme soul this cycle stops

On scientific analysis of this stanza this indicates time cycle. Time is governed by speed. Speed is circular or space where time exists which is also circular. Theory of relativity (of A. Einstein) the light ray released from one point will return to same point again. Travel of light has speed. Therefore, speed is circular. In fact, the soul (-Jeevatma) after death released from material body (a physical form), circular speed related to time, will certainly bring back to physical form of material existence with time. This is applicable to whole Brahm and material universe. Is it that the sages of Vedik time knew this theory of Einstein? Let us understand more from next stanza:

The nature has been described in two separate entities. One is manifest which is material world and the other is un-manifest. It is existing but nonmanifest form and is camouflaged. (GITA 16, 17, 18, 19). There is another un-manifest element. The un-manifest form of nature is eternal and transcendental to this manifest and un-manifested matter. It is supreme and un-annihilable. It is never annihilated. Manifest world is perishable (GITA chapter 8 stanza 20 given below).

Sahastrayugpryantmaharyad brahmano viduh I
ratrim yugsahastrantam tehoratravido janah II 17 II

(Gita 8, 17).

Day of Brahm is 1000 Yug so is night of Brahm.

Parastasmattu bhavonyovyaktovyaktatsanatanah I
ya sa sarveshu nashyatsu na vinashyati II 20 II

(Gita 8, 20).

It is interesting to note that the universe is recurring type i.e. manifest and annihilates again and again (see above punaravarti). In other word, it undergoes cycles of repeated birth and death. Gita 8, 16 given above.

To elaborate this, the original scientific explanation needs scrutinization. What is "yug" and "Brahm" mentioned in stanza 17 above? What is day and night of Brahma? During the days of BRAHMA all living beings of manifest nature come into being. During nights of BRAHMA all is annihilated. This is cyclical. It is repeatedly occurring during every day and night of BRAHMA. But when life of BRAHMA ends, they are all annihilated and remain un-manifest for millions of years. When BRAHMA is born again, they all manifest again after Brahm. How this was made knowledgeable? It is enigmatic. It is unbelievable. (See 7th stanza of Nasadiya sukta of Rigved above) These achievements are not possible in one's lifetime. Before understanding the life and time – period of Brahma. Understanding of the celestial movements in space – ANTARIKSHA. Vedik rishis understood celestial drama in space. The explanation is given in SHIV PURAN. To start with, assume that unmanifest element-SHIV self-manifest, as AGNI-LING, Having tremendous heat. Gasses are emitted which surround AGNI LING. Swirling balls of gasses are ejected out. The gas balls once thrown out of hot agni ling, the balls, keep rotating and after some years of cooling get solidified and become celestial objects which become galaxies full of different stars. These keep on rotating around central star of Galaxy. Some convert into solar systems with different planets. These also rotate around the Sun like stars. In our solar system the planet earth rotates around the central sun. This rotation decides time on earth i.e. Day, Night, month, and year etc. Similarly, the total mass of galaxies rotates, around Gas accumulation around AGNI LING. As explained in earlier sections of this book, that the Laws of universe are intricate and difficult to understand but have uniform applicability. The rotational celestial drama is seen in small atom also where electron and proton in atom rotate around the central neutron (explained earlier with diagrams).

What is the measurement of time? Modern quantum theory of science understands that the electron is just a charged entity. It is not a matter. It exists because it is in motion, rotating in its orbit. The existence of electron is in its motion or speed. Existence of electron

is because it moves around oppositely charged Proton. Hence attract each other. Mutual attraction of both, reduces, their charge. The time for spin of electron is RENU Sanskrit terminology, is its protection too is TRASRENU (TRAS = Protection and RENU is electron, (see – anoraniyan mahtomahiyan). The cosmos is filled with wavelengths of different types, depending upon expression of existence. Time required for electron wavelength is same as needed for energy to travel to infinity. When electron orbits around proton with opposite charge then why electron continues moving around in the orbit. Here the slip theory (of Neil Bohr. A Danish physicist, Nobel Laureate 1922.) is applicable. Electron can jump from one orbit to another by absorbing or releasing energy. **Such electrons when unite in time, space, and relative number form compound, molecules, and the material world in universe. The number of electrons, in outer orbit determine the properties of element (Neil Bohr).**
(www.nobelprize.org>summary; www.britanica.com>physics; 1.m.wikiquote.org)

As per Vedik knowledge the Cosmos expansion is limitless and unending. there are innumerable worlds beyond our limited world. There is no space, no time, no existence, and there is nothing. What we experience is all illusion (see Yogvasishht above).

Let us understand all this celestial drama perceived by Rishis and sages of Sanskrit speaking civilization. What science is hidden in this statement? According to BRAHM PURAN SRISHTI KHAND, the age of BRAHM and time scale of YUG has been calculated and mentioned. BRAHM lives for 100 years. In other words, 100 years is life span of Brahma. This is known as PAR Half of this i.e. Brahma's 1/2 life i.e. 50 years of Brahma is called PARARDH (half of par = par+ardh). The **time scale** of Sanskrit is as follows:
(For details refer to "GITA Press Gorakhpur" publication – Shri Shivmahapuranank Uttarardh No. 1, Year 92 – Shri Shivmahapuran-Vayaviya Samhita Adhyaya 7 Shloka 1-25 and Adhyaya – 8, Shloka 1-31, page 450-58).

15 NIMESH = 1 KASHTHA (time needed for closing of upper eyelid = NIMESH)

30 KASHTHA	=	1 KALA
30 KALA	=	1 MUHURT
30 MUHURT	=	1 DIN-RAAT (day-night)
30 DIN-RAAT	=	1 MAAS (month) (like man month, "DEVTA month" also has 30 days, 12 months is 1 year but time-period is different than the earth (कालावधि)
1 MAAS	=	2 PAKSHA – (SHUKLA PAKSH and KRISHN PAKSH).
6 MAAS	=	1 AYAN (sun solstice, southern and northern – UTTARAYAN and DAKSHINAYAN)
2 AYAN	=	1 VARSHA (year).

("SHRI SHIV MAHAPURANANK Editorial note, page 26, "BRAHMA KI AAYU" (KAL KA PARIMAN & TRIDEV AAYUMAN)".)

What is YUG and what are various YUG's? What is the time span of each YUG? Time span on earth is grouped into four different YUG. These are named as SAT YUG, TRETA YUG, DWAPAR YUG and KALI YUG.

12000 years comprise 4 YUG (YUG CHATUSHK) i.e. SATYA YUG, TRETA YUG, DWAPAR YUG, KALI YUG, all 4 together. As per calculation the number of years in each YUG respectively is 4, 3, 2, 1 DIVYA years. At the end and beginning of every YUG there is joint time – a transition period (transition of one yug into next yug) which is 100 years. All 4 YUG together are called CHATUR YUG.

One thousand CHATUR YUG is equal to one day of BRAHMA. Details are as follows:

(As per BRAHMA PURAN Gita Press Gorakhpur, code 44, SRISHTIKHAND page 20-22.).

Satya Yug time span is 4000 Divya years. In the beginning 400 years and at the end are 400 years. (800 years – Two joint periods of 400 years each, is the transition period between two Yug called Yugsandhi). The period of duration of Satya Yug is 4800 Divya years. Like this Treta Yug is 3600 Divya years, Dwapr Yug 2400 years and Kali Yug is

1200 Divya years. (See Manusamhita stanzas scripted below). Manu a great sage – illumined Rishi – of Satya Yug, has written more clearly about four Yug in his Samhita: Years 4000 is Satya Yug, Twilight of morning and evening (Yug Sandhya) joint (Sandhi) period, 400 each, make 4800years. The thousands and hundreds decrease by one (3000 +300 and 300 =3600 Treta yug), 2000+200+200=2400 is Dwapar yug, 1000+100+100=1200 is Kaliyug) Total is 12000 years is Devyug. The 1000 years of Dev Yug is a Day and equal is Night of BRAHM. It is described in Manu Samhita:

Chatvaryahuh sahastrani varshanantu krutam yugam I
tasya tavachchhati Sandhya sandhynshch tathavidhah II

itareshu sasandhyeshu sasandhyansheshucha trishu I
ekapayen vartante sahastrani shatani cha II

yadetat parisam khyatmadavev chturyugam I
etad dvadashasahastradevanam yugamuchchate II

deivikanam yuganantu sahastram parisankhyaya I
brahmamekamahardnyer yam tavati ratrirev cha II

(DIVYA year is 360 times more than the man years).
(Quoted from "The holy Science" by Sri Sri Swamy Sri Yukteshwar Giri, page 11, Yoganand Satsanga Society of India. Seventh Impression, 2013. Distributed by Jaico Publishing House).

Accordingly, duration of Kali Yug is 4,32,000 years. Similarly, Dwapar Yug is 8,64,000 years, Treta Yug is 12,96,000 years and Satya Yug is 17,28,000 years. The sum of all these is 43,20,000 years. This is one Chatur Yug. Like this 1000 Chtur Yug is one day of BRAHMA. That is 4,32,00,00,000 man-years is one day of BRAHMA. In one day of BRAHMA there are 14 Manvantars. 71 Chatur Yug make 994 but one day of BRAHMA is of 1000 Chatur Yug. Hence 6 Chatur Yug are remaining. If 71 Chatur Yug day of BRAHM counted then it makes 4,29,40,80,000. man years. If 14th part of six more Chtur Yug counted it makes 4,32,00,00,000 years which is day of BRAHMA. After day of BRAHM night starts. The night is time of annihilation of life of living beings. Then again day starts, and life begins. This cycle goes on till 100 years considered as life of BRAHM. **This is the meaning of GITA stanzas mentioned above.** Scientific aspect, as per modern science is, the galaxies are rotating around centre of universe. Solar system with all Planets rotates, around centre of the milky way galaxy. Similarly earth and all planets of sun rotate around sun.

Diagram to show EQUINOX movements "Piscean-Virgo" age (arrow.). Next equinox will be "Aquarion-Leo age. Transit of various yugas is in outer circle. Signs of zodiac in inner circle. Age is as per opposite signs of zodiac (Virgo-Pisces)

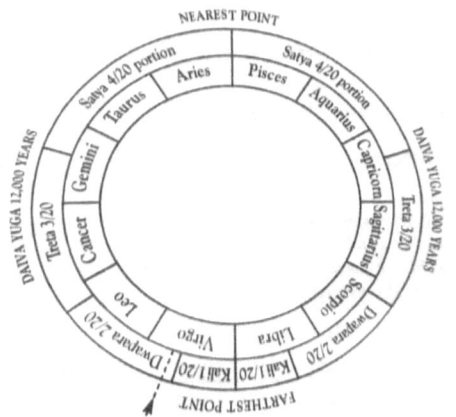

"(Diagram: Virgo is the sign opposite Pisces. The autumnal equinox is falling in Virgo the opposite point, the Vernal Equinox, is perforce now falling in Pisces – Piscean Age (arrow marked). With retrograde movement will enter Aquarius-Leo. According to Shri GIRI's Theory, the world entered the Pisces – Virgo age in 499 AD, and will enter Aquarius – Leo age in 2000 years later in the year 2499). Effect of such rotation is explained in text below

YUG description is nicely encrypted in MANU SAMHITA. The illumined RISHI says in his shlok as mentioned above. 4000 years is TRETA yug, Joint period between two yug beginning and end i.e. 400+400, Total 4800 years. Similarly other three yug In 1000 and in 100 reduce by one i.e. 3000+ 300+300 total 3600 years and so on. Four yug together i.e. 4800+3600+2400+1200, Total 12000 years. This is the age of divine Gods. Sum of thousand divine ages constitute one Day of Brahm. Equal is night of Brahm.

With reference to the diagram above, it can be summarized as follows (for details refer to original text):

According to Shri GIRI's Theory, the world entered the Pisces – Virgo age in 499 AD. and will enter Aquarius – Leo age in 2000 years later in the year 2499.

11501 B C when Autumnal equinox was on first point of Aries, the Sun began to move from nearest to centre, away towards point farthest from it. Autumnal equinox Aries – Libra the Sun moves away from the grand centre i.e. $4/20^{th}$ portion – Satya Yug. Human Intellectual power began to reduce. During, 4800 years, the intellect of man lost power of grasping spiritual knowledge.

During next 3600 years the Sun moves descending to next $3/4^{th}$ portion – Treta Yug. Divine magnetism reduces and in human minds.

During, 2400 years next Sun passed through Dwapar-Yug, human mind lost the power of grasping electricity and its attributes.

During 1200 years of Kali Yug transit of Sun, which is farthest from the grand centre, at the first point of Libra, the intellectual power diminished to only material greed and sense gratification only or 500 A D considered the darkest part of Kali Yug in the whole cycle of 24000 years. Widespread ignorance, sufferings in all the nations of the world.

From 499 A D onwards, Sun started to move towards centre, the human intellectual started to improve. During 1100 years of ascending Kali Yug, i.e. now 1599 A D. human intellect could comprehend fine matters of creation, Electricity. Next 100 years of joint period with Dwapar human intellect perceived attributes of electricity and fine matters of creation. The practical use of such knowledge like Magnetism, laws of astronomy, development of telescope, laws of gravitation etc. started advancing.

After 1899 on completion of 200 years of joint period of Kali – Dwapar yug, thorough understanding of attributes of electricity unfolds to human mind.

History is the testimony to all these events of recent past. This happens in a cyclical manner. It keeps on repeating. **This reiterates** the fact that invention of Battery – the cell was invented by Rishi Vasishtha and used in generating Ozon gas using Basil leaves (TULSI) the same was later invented by Nikola Tesla. Is it not a repetition of invention? This certifies the truthfulness of the diagram (its interpretation) and its authenticity.

Such is the influence of TIME-CYCLE (Samay Chakra) which governs universe. Comprehending the kingdom of supreme soul.

'Man' year has 365 days as earth rotates around sun in 365 days. Sun rotation around centre of galaxy takes certain time (one rotation 360 days = one solar year). Solar seasons also change every 30 solar days which is month of sun). Seasons of sun also change, as on earth i.e. every 2-4 month. Galaxy takes its own time to rotate around centre of universe. BRAHM takes its own time for one rotation. That is how the day, night, and year of BRAHM was calculated. **This knowledge was well known to sanskrit speaking civilization and it was common to all public. Therefore, this has been cited as example to arjun by Lord Krishna.** BRAHM also takes its time to rotate around Vishnu. Time period for Vishnu is also mentioned. The Vishnu revolves around Shiv. Period of rotation of Vishnu is also set as Vishnu rotate around Shiv. Details are narrated elsewhere.

(Adopted from "The holy Science" a work by JNANAVTAR SWAMY SRI YUKTESWAR GIRI. 8th Indian edition 1990 (7th impression 2013), Published in India by YOGODA SATSANGA SOCIETY OF INDIA. ISBN 978-81-89535-19-3.)

As per **Oriental Astronomy** moons of a planet rotate around their planets. Planets along with moons rotate around their Sun. SUN with planets and moons rotate around some star. It takes 24000 earth years in one rotation. Sun also has another rotation around VISHNU NABHI – A seat of creative power – BRAHM. As per chart sun takes 12000 years up and down. *The intellect of humans depends upon this rotational position. Changes take place both externally in material world and internally in intellectual world.* Reader is referred to original work for details.

Practical use of this in Sanskrit speaking civilization has also been established. While worshiping, honouring, reverencing "the GOD",

Nature called SHODSHOPCHAR PUJA i.e. 16 rituals/types of reverencing, the place position period of all the planets, stars, signs of zodiac, position of moon and many other planetary details etc. (details are out of scope of this book) are clearly pronounced to start the PUJA CEREMONY – where on earth and at what place the puja is performed. *This means the space science and earth geography is utilized in every walk of life.* In modern days while on a flight during air travel at present the airhostess/pilot, mentions the position of aero plane. Air plane's height, altitude, temperature, speed, distance from departed airport and remaining distance to destination of arrival at the airport. Is it not a practical use of knowledge? Similarly, that civilization used their knowledge in every walk of life. Some details are given below about scientific use of knowledge of space, planetary positions of stars and nakshatra, earth's position in space, position of Sun and moon and in which sign of zodiac sun, moon are present. Apart from this the geographical position of site of reverencing to super soul (Paramatma). Site of place of worship in relation to River in the region, all are taken into consideration before taking oath and determination for reverencing along with prayer to GOD GANESH – to take care of all the obstacles and problems which may arise during reverencing. Instead of appreciating science in this it is propagated as orthodox and un-necessary repetitions. A question comes to mind why on flights the position of aircraft is repeted again and again. Is it not orthodox? Probably the idea was to condemn whatever is BHARATIYA and make Bharatiya (Indians) feel inferior. So that the invaders are superior, and science oriented. The Reverencing is cited as for example:

16 types of reverencing, **Primordial BRAHM,** Second half of BRAHM, **BRAHMA'S 2nd half life of BRAHMA**), After Brahma position of VISHNU i.e. KALP named SHWETVARAH), MANU named VAIVASWAT **MANVANTAR (one of 14** manvantar)**, 28th out of 71 YUGCHATUSHK, group of four yugs (Ashtovinshatitame yug chatushke) First leg of YUG named KALI,** JAMBU named continent

Indian subcontinent, Deccon (Dakshin pathe) abode of RAM, **Period of BUDDHA avtar and so on – DANDAK named jungle region AARANYA,** south bank of river Godavary (SUN's particular sign of zodiac on the day of puja, moon's particular sign of zodiac on the day of puja, In fact position of earth in space and other planets, stars, NAKSHTRA (27 IN NUMBER. Details are in ephemeris – PANCHANG-celestial navigation calendar), PLACEMENT OF Sun, Moon and their position in which particular sign of zodiac is explained. This narrates the position of earth on puja day in whole universe. Even to-day, this is in practice while performing PUJA CEREMONY. All sounds highly scientific. Condemning such scientific narration are we not condemning science itself!!

This gives an understanding that every citizen of Sanskrit speaking civilization was aware of knowledge of space, astronomy and astrology and Ephemeris and geography of the place on earth). The scientific knowledge of the civilization cannot be ignored. **It was very well advanced in many respects, in comparison to modern space age knowledge.** The occurrence of ECLIPSES of moon and sun and that too accurately, was routinely calculated and scripted in advance for the whole year. It is going on year after year even now (in PANCHANG published every year). The names of HINDU months are according to the presence of moon in that NAKSHTRA viz. When moon in CHITRA NAKSHTRA the month is CHAITRA. When moon in VISHAKHA NAKSHTRA it is VAISHAKH and so on JAISHTH-JESHTHA, etc. In summery all derivations are highly scientific. Not only this, but the effect of the celestial phenomenon in space affect all humans of the world, was also established by ancient RISHI's of Sanskrit speaking civilization. This is all available in ASTROLOGY and ASTROGNOMY developed by the *rishi muni and saints*. In fact, it is not an exaggeration to say that these highlighted words are synonyms of Scientists of Modern civilization. All astrological derivations are true even today. The calculations are scripted down and available even today. It is available in some centers in India. One such center is in

Hyderabad, Andhra Pradesh. Which is known to author of this book. It is called NAADI SHASTRA. (At this "center" the organizers take thumb impression. Which is compared with the script available and if it matches your past and future is just read out for you.). It can be verified by any body by visiting the center.

How scientific, intellectual and political scenario is influenced on earth due to celestial drama in space has been charted out by these RISHI and sages of Sanskrit speaking civilization.

The intellectual development, scientific development and political scenario of nations of the world, all in fact, due to the **celestial phenomenon in space.** The first 12000 years (see diagram, page 208) are responsible for complete change in external material world and internal intellectual mental level of mankind. 12000 years are divided into 4 different stages. The first 1200 years, sun transit during $1/20^{th}$ portion of orbit is KALI YUG, during this period the mental virtues are $1/4^{th}$ developed. Only gross material world is acceptable and understood. During next 2400 years, sun transits through $2/20^{th}$ period of its orbit, known as DWAPAR YUG, mental values are in 2nd stage. Human mind perceives fine matters and electricity which form external world. The period of 3600 years sun transits through $3/20^{th}$ orbit, is TRETA YUG. The mental values are in third stage. Human intellect can now comprehend divine magnetism, source of all electrical forces on which creation depends. The period of 4800 years sun transit is in remaining $4/20^{th}$ orbit, is SATYA YUG, mental values are in 4th stage, when human mind and intellect now comprehend all even GOD – spirit, which is beyond material world. (For details reader is advised to consult original literature, as it is beyond the scope of this book).

The un-manifest nature on the other hand is eternal and transcendental to manifest nature. It is supreme and is never annihilated. When all world is annihilated that part remains as it is. *Interestingly after narrating all such knowledge Lord Krishna said all universe is perishable in cyclical manner (i.e.-rotation), then why ARJUN is worried for such a perishable beings. Lord Krishn convincing ARJUN to follow his own KSHATRIYA DHARM and leave the outcome to super soul.*

During such conviction many scientific facts are mentioned as examples. Attempt is made to explain such hidden fact by original work found in various available and relevant Sanskrit literature on the subjects.

The yug calculations are based on scientific calculation principles. These principles and calculations were prevalent in BHARATVARSHA till the reign of king VIKRAMADITYA. After this it was introduced as SAMVAT and named as VIKRAM SAMVAT. The days of week are named after planets as Monday – SOMWAR = MOON, Tuesday – MANGALWAR = Mars, Wednesday Budhvaar – on planet Budh = Mercury-BUDH GRAHA and so on. Similarly, Months are as per NAKSHATRA movements as stated above. CHAITRA is as per the CHITRA NAKSHTRA, VAISHAKH as per VISHAKHA NAKSHTRA and so on.

One more inference can be derived that the knowledge of ZERO is very very old. Because of this zero, man could calculate accurately the calculation of stars and planets and it was made easy. The discovery of zero was a necessity.

Pious literature or Shastra totally depends upon Mathematics. Without zero, calculation in maths is difficult. In Brahmotpatti the existence of manifest world came into being. Self-manifest SHIV, is basically creator of all creation and creature in universe. Shiv is responsible for the Primordial Gasses in the form of Vishnu. Hot primordial Gaseous balls from the center of Vishnu in the form of Brahm, which is responsible for different and innumerable Galaxies which came into existence. Manifestation of main elements like Earth, Air, Fire, Water, and Space as per Gita 7, stanza 4 and 5, (Chapter 7,) were possible by Brahm. In Prajotpatti with RAYI and Pranik energy living beings came into existence. Both conglomerate and responsible for galactic manifestations. The celestial drama is going on since time immemorial. To know the effect of "TIME

CYCLE" on living beings the time was graduated by Sages, Rishis or contemporary in hours, Days, weeks, and months etc. and given names as per the Planets, Nakshatra etc mentioned above, also are under the influence of the cosmic manifestations. To study the influence of these cosmos was analysed and grouped and to study their influence on time and living beings.

This is what Lord Krishn says in Gita Chapter 9 stanza 7, – *"kalpkshaye punastani kalpadou visrujamyaham\"*, At the end of previous Kalp and beginning of next kalp, nature manifest the universe again.

Chapter 10, Stanza 21, – *"aadityanamham vishnurjyotisham ravviramshumana"*, Among 12 adityas 'I' am by the name 'Vishnu' and jyoti illumined one fraction is in Sun (Ravi).

Chapter 10, Stanza 30 –*"pralhadshchasmi deityanamkalah kalayatamaham\"*, 'I' am Prahlad (the devoted one) and death for demons, evil in the societies.

Chapter 10 stanza 33 –*"ahamevakshayah kalo dhatamaham vishvtomukha:"*, 'I' am unending time (time is the unending and ultimate killer) and creator of universe.

Chapter10, stanza 35-*"masanam margshirshoham rutunam kusumakara"*. 'I' am month Margshirsha (9th month or 'fall' season when tree leaves ripen and change colour, before shed off). New spring season comes is "ritu kusumakar". These changes are natural manifestations due to supreme soul effect.

Life span of Brahm, life of Vishnu, and then life of Shiv has been explained in Markandeya Puran.

Ashtoyugsahastrani ahoratra prajapate aneinevatumanen shatambrahmasi jeevati pitamaha shateneiva Vishnorman vidhiyate nimishardhenshambhostu sahastranichaturdash vinashyanti tatha vishnorsankhyatahpitamaha.

8000 yug is Day and Night of Brahma, accordingly life of Brahma is 100 years. During span of ½ NIMESH of SHIV, 14000 VISHNU are included. In a day of Brahma there are 14 Manvantar. 71 ChaturYug is one Manvantar. 30,67,20,000 Man/earth years is one Manvantar or 8,52,000 Divya years is one Manvantar. Multiplied by 14 manu is 1,19,28,000. This is a day of Brahma. In this manner 4,29,40,00,000 earth years is one day of Brahm. Like this, 100 years is life of Brahm. As said earlier everything manifests and continues during day of Brahm.

Chaturyuganam sankhyata sadhukahyeksaptatihi I
manvantaram tasya sankhya manushabdeirnibodhame II 34 II

Trinshat kotyastu sampurnah sankhyatah sankhyayadvij I
Saptashashtistatha nyaniniyutani cha sankhyaya II 35 II

Vinshatishcha sahastrani kaloyam sadhikamvina I
etan manvantaram proktam deivyervarshernibodhame II 36 II

Ashtouvarsha sahastrani divyaya sankhyaya yutam I
dvipanchashattathanyani sahastranyadhikanitu II 37 II

Chaturdash gunohyeshkalo brahmyamah smrutam I
tasyante pralayah prokto brahmo neimittikobudhei II 38 II

Markandeya Puran Ch. 43

(Markandeya Puran-chapter. 43 (brahmaji kia ayu ka pariman) stanza 8 Maarkandeya puran 145).

As stated above Gita in chapter 8, stanza 16, 17, 18 Time needed in Prajotpatti (explained earlier), manifest world repeatedly appears and disappears. Similarly, the knowledge, research and development **is also,** in a cyclical manner. Science also reveals to humans in cyclical manner. This is exactly mentioned in the diagram quoted above. (Diagram in chapter 8 page 208).

Modern science has now come out with theory that the manifest universe of ours has originated from an unknown source which Gita recognizes as (avyakta) Brahm. Recognition of black hole where existence is totally devoured. What happens to those existing materials in black hole, is beyond knowledge. Everything reaches to nothingness.

Star, suns, galaxies etc. all are condensed to nothingness. Then again, the black hole explodes and everything manifests again is the opinion of some scientists. This is the cycle of evolution and destruction. Existence is always with light and destruction is always in the darkness. This is what Gita believes.

Bhutgramh evayam bhutava bhutva praliyate I

ratryogamevash parth prabhavatyaharagame II19 II

(Gita 8, 19).

All living beings repeatedly manifest during day (light) of Brahma and annihilated back to Brahm during night (Darkness) of Brahm. But the supreme remains unchanged. For super stage see Nasdiya Sukta as above.

In summary, the truth is in creativity and its creation of light energy which is intermittent through positive rotation of electron. When rotation of electron is negative, light energy ceases. Electron can jump from one orbit to another by absorbing or releasing energy. **Such electrons when unite in time, space, and relative number form compound, molecules, and the material world in universe. The number of electrons, in outer orbit determine the properties of element (Neil Bohr).** All this has been discussed earlier in this chapter.

Purport of this can be concluded as follows:

The electron being energy, decides the rebirth (darkness or bad cultivations) or salvation (Moksh, light or good cultivations) of entrapped soul energy in material manifest body. The life-time cultivations during span of life, influence the electrons by altering the number of particles in electron (good or bad depending upon cultivations.) Accordingly influence the electron. The behavior pattern (swabhav) of individual also can change (discussed above). Without microlevel change behavior is difficult to change. Good or bad cultivations influence the microlevel of electrons orbiting protons, and alter the physical, material manifestations in universe. Time cycle has its own influence on all processes. It may take long time for the genetic transformations. Because of such knowledge the Sanskrit speaking sages and rishis implemented 16 sacraments in the society (Solah sanskar).

CHAPTER NINE

"RAJVIDYA RAJGUHYA YOG" SECRET, MYSTERIOUS, ETERNAL, TRUE AND CONFIDENTIAL KNOWLEDGE*

Abstract: Up till now, in previous chapters of GITA it has been convinced that the body and soul are separate entities. Body is perishable but soul is eternal. Here in this chapter Lord Krishna narrates to Arjun the most secret and confidential knowledge called as "RAJVIDYA RAJGUHYA YOG" obtained from royal wisdom. Writer of Gita seems much ahead of its time stating idealism for the upliftment of human societies of the world. Individual worship or propagators of faith, cult, religion transforming humans into sectorial beliefs and indulgence into worship of superman responsible for religious ego. I, my or mine sect, faith, religion is superior, better to follow and adopt. In Gita words like "I", "My", "Mine" are used for ultimate super energy or super soul. Lord Krishn in Gita is symbolic of super soul (absolute truth) in the form of incarnation and not individual superman. This has been stressed

---

* In fact, there is no perfect word for GOODH/GUHYA in knowledge of author therefore so many equivalent words used to represent Sanskrit word "param guhya dnyan".

in Gita repeatedly. Vedik tradition has no belief in individualism. Complete idealism of devotion towards perfection up to super reality or absolute truth. Knowledge be accompanied with scientific information, is the basis of Ved and vedik philosophy. It has been discussed earlier by use of words like "Na Iti" or "Neti" meaning there by – "thus so far" we know, further exploration be researched out by you and next generation. Followers of ved and Gita, must think for themselves. Idea behind is person must become perfect. Independently march towards liberation. Everyone and everything of material universe is with same substantial energy fraction of "super tejahpunj" – store house of unique super energy. Hence individuality has no role to play. Parity, love, bliss is opinion of 'yog cult'. Science or VIDNYAN is not only about technical aspect but, experience and wisdom, in fact liberation from miseries of material existence at the same time enjoyment of life to its fullest extent.

In (stanza 1), Shri Krishn mentions to Arjun:

Eidam tu te guhyatamam pravakshyamyanasuyave I
dnyanam vidnyansahitam yajdnyatva mokshyaseshubhat II 1 II
(Gita 9, 1).

Lord Krishn narrates that a confidential knowledge with scientific bend is here for humanity (in disguise as Arjun), which will certainly liberate everyone from miseries of material existence. Science indicate technical achievements but VIDNYAN means not only technical achievements, but knowledge and wisdom achieved with actual experience as well. This is contrary to numerical supremacy and fanaticism in faith with less knowledge and wisdom, which is far away from the welfare of the humanity.

Rajvidya rajguhyam pavitramidamuttamam I
pratyakshavgamam dharmyam susukham kartamvyayam II 2 II
(Gita 9, 2).

Royal wisdom and confidential royal secret knowledge, pure and best, provides happiness, and joy of everlasting nature. Understanding difference between body and soul is knowledge which is secret and confidential. This transcendental knowledge involves soul and supreme soul of manifest and un-manifest nature of the universe. This knowledge is essence of Science of spirit.

Nonbelievers, in this understanding remain in darkness of ignorance and involved in cycle of birth and death Stanza 3. Such knowledge is not a curriculum in any of the universities in the world. The soul is the root cause of the body and is ignored because majority of people are engrossed in sense enjoyment, sense gratification and involved in gain in material world due to camouflaging effect of the material world i.e. MAYA. Why this subject has not been introduced in schools, colleges and universities? It is a big surprise! Even though it is important for tranquil life. It relieves man from stress strain of material world. It is needed in modern times more than the past because the society is under lot of stress and strain. More and more stress related diseases are seen in modern times. The knowledge will certainly produce serene, thoughtful and perfect human being, and in turn better human society will be an outcome. It will give ego-less citizens to countries and promote universal brotherhood. It will help in achieving the vedik concept and the goal of – whole world a family.

If there is no soul/energy/spirit in the body, the body is lifeless. In absence of this soul/pranik – energy in the body, it is dead. From where this energy comes and where it goes it is a speculation. In "ASHWAMEDH PARV" chapter of MADBHAGWAT GITA, (post Mahabharat war period) ARJUN requested Lord KRISHNA to repeat all what was narrated on the day of beginning of MAHABHARAT war. But KRISHNA could not repeat it. HE was in a trance of meditation on that day and was not able to repeat again the whole statement narrated that day. This signifies that the deep true knowledge only possible under transcendental meditation only. Once that meditative "trans" is over, that thought tends to disappear. Whatever memory can be regained and "trans thought" scripted down it is for the material world. That is why the VED are called superhuman knowledge, as stated previously. This is from the experience of the author during research on regenerating organs and tissues in the body vide international patent from USA. One can ask that why the spiritual knowledge is important to living being or man. Engrossment in material world never lets one to understand the true, deep, confidential knowledge. It is camouflaged due to greed and sense enjoyment of material world. Human minds are lured towards material world due to camouflaging effect of material nature (MAYA).

Pertinent questions remain unsolved are: Why living beings are born? Once born what is the purpose of birth? Should one live and die for sense enjoyment only, in between birth and death? What is our aim in living life? Is there anything beyond life? Is there super soul or energy in this universe? Who governs or controls life and the universe as a whole?

Partially this has been covered in previous chapters. In an attempt, to solve such questions Sanskrit literature, knowledge in VED and PURAN, the spiritual existence has been discussed. Apart from material knowledge, Ved could achieve, scientific knowledge about space, astrology, mathematics, medicine, physics, chemistry etc. Along with this, spiritual knowledge could be acquired. The Sanskrit speaking civilization of Vedik period, achieved occult and mystic knowledge as well. Not only knowledge about material world the spiritual knowledge was also well developed. VED, PURAN, are a testimony to this. These two knowledges are compared previously (Chapter 2, page 80).

After achieving, all the material knowledge, it was thought that the spiritual knowledge is ultimate and true and confidential. This knowledge is most difficult to achieve. Engrossed in material world and involvement in material existence, eludes the true knowledge due to camouflaging effect of this material world (MAYA). In BHAKTI YOG – devotional YOG, mind is focused on one subject, (god, creator, BRAHMA etc.), hence yield is obviously and logically more (as discussed in 2nd chapter). Energy used by brain is not dissipated. In other word, the action is free from stress and strain of life. It yields, maximum output from minimum use of energy and action. **In wakeful silence, with**

minimum efforts and use of maximum potential of intelligence, a lot more can be achieved. This can be compared with the speed and fuel consumption of an auto-vehicle. Uniform speed without any ups and down or upheavals, the fuel consumption is minimum and yields more benefits. Therefore, this quality be developed in every human being for understanding laws of nature and their utility. This is the preaching of VED and PURAN of Sanskrit speaking civilization. This is upheld by RIGVED and is also the approach of modern science like Physics, Chemistry, Mathematics, Biology, Cosmic quantum science.

As per stanza of Gita in the beginning of this chapter, Gita Ch 9, 1 and 2, the knowledge is considered everlasting and is most supreme or KING EDUCATION, KING (as per Sanskrit stanza) confidential knowledge, purest of pure knowledge. It develops the perception of self-realization. It is essence of all knowledges. Transcendental knowledge gives understanding difference between body and soul.

Following such a statement about VIDYA all focus was probably on spiritualism after devastating effect of Mahabharat war. There was a great loss of population on both sides of warring forces. Peace was of utmost priority and all defences shattered during post war period. All focus on spiritualism. AVIDYA was ignored. *No one thought about the stanza of ISHVASYOPNISHAD (Chapter2) i.e. both physical and spiritual knowledges have equal importance for material body and material world.* During modern times, universally a great stress is placed on necessities of body and not the soul. But the fact is without the soul the body is nothing. One may think that once the soul leaves body it becomes inactive. The knowledge of soul and body explains that soul is eternally active. Its activities are in spiritual kingdom. Even if soul leaves body, its association with body cannot be disconnected immediately. Sinful or pious actions of body do remain associated with soul (explained in previous chapter). Fruits of the deeds and actions of body get carried forward with soul and to be enjoyed or suffered by next body/incarnation. In the same context a verse from PADMA PURAN:

Aprarabdhphalam papam kutam beejam phalonmukham I
krameneiva praliyate vishnubhaktirtatmanam
(Padmapuran).

An example of seed is used to explain this in PADMAPURAN as well. Fruits of actions good or evil, take time to show effect as seed takes time to start germination so the fruits of actions. Those who have devotion to supreme soul are devoid of fruitful reactions of deeds-good or bad. Similar statement is in 7th chapter already discussed.

Yesham tvantgatam papam jananam punyakarmanam I
te dvandvamoham nirmukta bhajante mam dradhavrata
(Gita 7. 28).

The entire universe is pervaded by the super soul but not the vice versa. Whatever is manifest universally the super soul is not a part of this cosmic manifestation, though HE (super soul) is everywhere, because HE is the source of all creations. The automation in cosmic manifestation, created and annihilated again and again is totally under control of super soul, but super soul remains detached from all this cosmic activity. HE is the eternal seed of all universe.

The relation between cosmos and living being (in universe everything, manifest or un-manifest is living – having birth, growth, change, time span, and death) is expressed in a statement: "Anoraniyan mahatomahiyan". The Cosmic laws are complete whether the expression is smallest of small or biggest of big. Super soul is omnipresent and omniscient. HE is fully awake about nature and natural laws. Total study of laws of nature expressed and unexpressed universe is common basis of everything and everyone. Same fact has been narrated in vedik script: (Yatha pinde tatha brahmande, and Aham bramhasmin).

As is the atom so the universe, similarly, as is the body so is the cosmic body. Laws of the universe are camouflaged by the revealed world but have uniform applicability. To explain this, it is well known that earth along with other planets rotates around sun. Sun with solar system rotate around center of the galaxy. Galaxy rotates around center of the universe. Smallest particle on earth the atom has electron and proton rotate around central nucleus. The earth has $3/4^{th}$ water and ¼th land so is our body and smallest unit of our body cell has similar composition. This has been explained earlier in section of General considerations.

Aham brahmasmin is another statement from VEDIK literature. This also states the same as above. Scientific aspect of all this is in everyday life. Whatever is happening in universe is also present in the revealed/expressed world of universal consciousness.

Author as medical professional has observed in his clinical practice that patients with diseases of similar nature and organs, come in crops i.e. diseases of THYROID gland in a particular time come together in groups and similarly diseases with Prostate, Pancreas, liver, spleen etc. mostly come in groups. This can be explained by the astrological sciences which was highly developed by sages, RISHI-MUNI'S of Sanskrit speaking civilization. In Sanskrit literature there are 27 NAKSHATRA i.e. groups of stars in a "Unique and particular formations" called NAKSHTRA. These constellations have been given names. ASHWINI, BHARANI, KRITIKA, ROHINI and so on. These individual NAKSHTRA have influence on chromosomes/DNA of cells. These also control various glands and organs of the body. These NAKSHTRA rotate in space. Whenever a "particular NAKSHTRA" directly aspect towards earth and exert influence on a particular gland adversely the patients with diseases of that gland or organ come to clinics. That is how diseases are seen in groups. (This research was undertaken by author during 1970 – 1973 as Registrar in Surgery Department, of M. A. Medical college, New Delhi. But could not be completed due to transfer to Andaman and Nicobar Islands under Central Health Scheme of government of India, while in government service, follow up of patients was impossible. Preparation of horoscope was utmost difficult due to incorrect date of birth of patients. "Palm prints" to study palmistry and health, was lost. The palm is replica of horoscope and can be correlated by experts). This all needs research in collaboration with astrologers and clinicians. It is a subject of multi-centric and collaborative research. Interestingly MAHARSHI MAHESH YOGI has stated influence of stars, Constellations and 'Nakshtra' from space, on BRAIN, CELLS etc. as per Vedik Literature. He has discussed influence of cosmos and twelve signs of ZODIAC on DNA of cells of body Systems.

(MAHARISHI forum of Natural law and National law for doctors. 1994 MAHARISHI year of discovery of VEDA and VEDIK literature in human physiology. 1995 MAHARISHI'S year of silence).

Rigved statement emphasizes similarly "Brahma bhavati sarathihi" i.e. creator of material world guides universe. SARATHI means a charioteer i.e. driver of chariot who guides path of chariot, so BRAHM guides everything in material world. The above expression is a scientific derivation that through the effect of NAKSHTRA, Constellations and stars etc. in space, material world is under influence and under guidance of BRAHMA (see chapter 15 also).

In this chapter, Lord Krishn re-iterates to ARJUN, simply knowing that body and soul are separate, is not enough to know about soul.

Antavanta ime deha nityasoktah sharirinah I
anashino prameyasya tasmadyudhasva bharat II 18 II
(Gita Ch.2, 18)

Eternal soul in perishable body, never dies and after death it is liberated, is a common knowledge. It is not a factual knowledge. How a soul which is in association and active with body for such a long time be free? It is active. This is the confidential part of the knowledge. As stated above and as written in PADMA PURAN, the fruits of sinful actions are nullified if a person is engaged in devotional service to Lord – (supreme soul). Otherwise the sinful actions and its fruits are carried forward and sooner or later are effectively executed in this or next incarnation. Regarding liberation or re – incarnation GITA has expressed its views repeatedly. Death is inevitable but liberation is not. It must be achieved and perfected by the individual. Generally, the people are educated in material knowledges, all over the world in different universities. In the universities of Sanskrit speaking civilization of ancient Bharatvarsha, now India, both subjects i.e. VIDYA and AVIDYA or physical and spiritual were the part of curriculum. Prominent teachers/philosophers were many. To name a few Gautam (NYAY DARSHAN), Kanaad (Vaisheshik Darshan-Particle physics) Kapil (SANKHYA considered mool prakriti independent causative element of manifest world while soul is cause of all causes), Yagyavalkya, Shandilya Vaisvanar etc. were the authorities on their subjects.

The work performed by the body – good or evil has its own effect. VEDIK literature propagates the theory of re-incarnation. No proof could be traced. Reincarnation of soul depends upon the work performed and its fruits, Soul is with the body but is not entangled by the work done by the body. Then how the soul is responsible for actions, need some explanation. Soul and body are united but not dissolved. The soul is involved in various bodily functions and associated with the body for a long time. A perfect example described is of magnet.

Nirichchhe sansthite ratne yatha lohah pravartate I
sattamatrena deventatha chayam jagajjanah II

Ata aatmani kartrutvamkartutvam cha sansthitam I
nirichchhtvadkarttasou kartta samnidhimatratah II
(Sankhya pravachanbhasya I, 97).

Without wish the Iron adopts dynamic movement just with the presence of a Magnet nearby, so is the body due to presence of the SOUL. Even though it is nonactive but mere by presence it becomes active.

Scientifically it can be explained by the following:

Body is composed of material elements from material world – Major material elements with Mind, Intellect, and ego stated above. It has Macro body, Micro body, and AATMA the –soul. According to TEITTARIYA UPANISHAD five subtle bodies described (**FOOD** body, **PRAN** body, – earth related, both can move on earth, **Mind** body – air related, can move in air, **Science** body – related beyond air (मह:), can move beyond air, **Pleasure** body, related to BRAHM, can move limitless). Out of these, the micro body is capable of leaving macro food body and can travel in the universe, at the same time, can return back to the same body. This has been proved by ancient rishi-muni of the past. In recent time, the same has been proved by Dr. P.V. Vartak in his book.

("UPNISHADANCHE VIDNYAN NISHTH NIRUPAN" 7TH Ed. 22-12-2018. Vartak Prakashan, page 8-15.)

He has proved his visit to Mars. His visit to Mars 10 days before the VIKING-1, visit in 1975 by NASA, USA. (It was to land on Mars and send information 11 months after launch from earth on 21st August, 1975. With the help of transcendental meditation and SAMADHI, Dr. Vartak's microbody reached Mars. And on return whatever he could recollect, he published his experience of mars visit in SANT-KRIPA magazine in 21 June, 1976. His description was almost 100% correct as per information of NASA about Mars exploration. He later visited again and seen the small VIKING-1 and 2, on Mars. Their union and separation as per orders from earth. He published his report on 13th August, 1976 in TARUN BHARAT, SAKAL, KESARI. On 18th August 1976, he reported this experiment to American Embassy, Bombay. Chief of embassy Mr. Bias appreciated the experiment and the information. Similar experiment he performed on Jupiter. Interestingly TEITTARIYA UPANISHAD has expressed in following Stanza:

Taseisha eva sharir atma I

yah purvasya I

tasmadva etasmadivadnyanmayat I

Anyontar aatma nandmayah I

teneisha I

purnah I

sa va esh purushavidha eva I

Tasya purushvidhatam I
anvayam purushvidhah I
tasya priyamev shirah I

Modo dakshinah pkshah I
Pramod uttarah pakshah I
aanand aatma I

Brahma puchchha pratishtha I
tadapyesh shloko bhavati II

(TEITTARIYA UPANISHAD)

Apart from this Patanjali has described six types of bodies. A human posseses different bodies or cocoons (KOSH). Their names are:

1. **Food body** (or ANNMAY KOSH). Body consisted with Earth, Water, Air, Fire and Space, accompanying soul energy.
2. **AIR (PRAN) body** (PRAN MAY Kosh). Body of five pran (panch pran) and action senses (five karmendriya). Described earlier.
3. **Mind body** (MANOMAY KOSH), Body of five senses and mind (mann). Described earlier.
4. **Science body** or Analytic body (VIDNYANMAY KOSH), body of intellect and ego.
5. **Pleasure body** (AANANDMAY KOSH). Body of Mind-set and accompanying soul
6. The ultimate sixth body is pure soul (Pure Aatm tatva), the Devine light. (DNYAN PRAKASH – Aatma – Jyoti).

The above stanza, from TEITTARIYA UPNISHAD explains that the body has soul. It is separate than "Pranmay" body. It is in fact an accompanying of mind body or Manomay body. The soul of food body is mind body. "I" in food body is PRAN body, but other separate body is mind body – an all knowing – OMNISCIENT. It is "micro" in comparison to air body, like invisible magnetic or electric waves. Inside this is intellect or science body. This body is knowledgeable body having special knowledge. This special knowledge is with even inexperienced child. Viz. Child by instinct knows that if he falls from top of a building he will die and hence he with instant reflex, steps back. He has no knowledge or experience in such an event. How he knows this? It is this knowledgeable "science body" having all knowledge is responsible for such an act on child's behavior. Death fear is an instinctive knowledge which is automatically is because of known to all. Such instinctive knowledge is due to the VIDNYANMAY KOSH.

Pleasure body or Anandmay body is micro aspect of soul. It is related to BRAHM, which is limitless. Hence its movement is also limitless. The five parts of body starting from Pranmay body, are in microform in comparison to other. The capacity of all bodies to move and limit to move has been explained earlier.

The manifest universe is totally under control of Super soul. The control is executed through nature and in turn by Stars, Nakshatra, galaxies etc. The creation and annihilation of manifest universe repeatedly is at the will of super soul. It is under automation:

Prakritim swamvashtabhya visrijami punah punah I
bhutgramimam kritsnamvasham prakritervashat II 8 II
(Gita 9, 8).

Incorporation of super soul with manifest nature, the creation gets manifest again and again. The living entities too get manifest with cosmic evolution and under influence of nature automatically.

Fear and dismay of Super soul, bourn by basic elements or Devta (Panch mahabhut) is responsible for orderly behavior and natural happenings in manifest material world. This is expressed in following stanzas of TAITTARIYOPNISHAD:

Yada hyeveish etasminnudarmantaram kurute I
atha tasya bhayam bhavati I
tatvev bhayam vidushomanyanasya I
tadapyasha shloko bhavati II 4 II
(TAITTARIYA Upnishad 2. 7. 4.)

Bhishasmadvatah pavate I
bhishadeti suryah I
bhisha smadagnishchendrashcha I
mrityurdhavati pancham iti II I II

2. 8. 1

Similar discussion is in BRIHADARANYAK UPNISHAD between Rishi YAGYAVALK and GARGI. When Sage GARGI (A LADY SAGE), asked that who is beyond main five elements and in whom all the manifest material world rest, in reply Yagyavalk explains that:

Etasya va aksharasya prashasane gargi suryachandramaso
vidhritou tishthat etasya va aksharasya prashasane gargi
dyavaprithivyou horatranyardhamasa masa rutavah sanvatsara
iti vidhritastishthntyetasya va aksharasya prashasane gargi
prachyonyandayah syadante shvetebhyah parvatebhyah
pratichonya yam yam ch dishamanvetasya va aksharasya
prashasane gargi dadto manushyah pash >\ santi yajamanam deva
darvi pitaronvayatah II 9 II
BRIHADARANYAK Upnishad (अं 3. ब्रां 8. श्लों 9).

All is under administration of Akshar (-unending)-the super soul in the revealed universe. Sun, moon and earth, time cycle etc. are in order under HIS administration. *Rigved statement emphasizes similarly Brahma bhavti Sarathi i.e. creator of material world guides universe.* This is the scientific meaning of the stanzas 7 and 8 of Chapter 9 in Gita (mentioned above).

The Prakriti or Nature indicated in these stanzas (7and 8 of Ch. 9) is scientifically analysed in Patanjal yog Sutra. To understand this one must clarify what is root or original element and what is Chetan or energizing element. Root element has two parts, formation and deformation There are 8 types of PRAKRITI as per Sanskrit literature. Basic Nature (Mool Prakriti), Important element (Mahattatva), Ego (Ahankar), with Five sense elements make 8 Prakriti – Sound, Touch, Sight, Juice, Smell are senses and their perception organs – Tanmatras). Accordingly, their perception by five sense organs (Tanmatra). Five 'sthoolbhut' or macro elements viz. Space (Aakash), Air (VAYU), Fire (Agni), Water (jal), Earth (Prithvi) and eleven senses (five sense – Indriya, Five work or Karmendriya) are total 16 deformation or Vikriti:-

Ashtou prakrutayah shodash vikarah purushah

(8+16+1 = Total 25)

Mulaprakritiravikrutirmahadadyah prakritivikrutayah sapta I shodashkastu vikaro na prakrutirna vikrutihi purushah II 3 II

(Sankhya karini).

In conclusion, this is the explanation about scientific aspect of Nature or Prakriti.

In this chapter one thing which has prominently come to understanding, is about whom to worship, formless super soul i.e. (Nirgun, Nirakar) or idol or deity with form (demigod). First one is difficult while the other is easy and can be perfected easily.

Scientifically oriented Sanskrit speaking civilization realized the fact that human mind perceives, mental image and form easily by majority of the population and hence IDOL worship was propagated in the societies. Devotion to formless super soul and concentration on formless GOD or zero super soul (Shunya brahm) is physically difficult. One in million hardly can perceive or follow this type of devotion. It is hard to practice prayer or meditation. The religions, cult or faith who believe on formless god are practicing on images like-Buddha, Christ, Mohammad, Guru Nanak, Mahaveer etc. Worship of or devotion to form, may be the first step to go for the formless devotion. Gita stanzas 13 – 26 of 9th chapter emphasizes, this principle. Lord Krishn reiterates that who so ever worships deity, demigod, form etc. ultimately reaches Super soul. Both the techniques are ultimately the same. In stanza 13 of this chapter:

Mahatmanastu mam parth deivim prakrutimashritah I
bhajantyananyamanaso dnyatva bhutadimvyayam II13 II

(Gita 9, 13).

Great souls depend upon and take shelter of divine nature mentally, and continuously desire for oneness with super soul, knowing it is a creation which is inexhaustible. Here BHAJ-, means devotion with emotion in mind (BHAV)-, i.e. tendency to become the same whom one is worshiping or DHYAN and meditation on the same. It should not be showy. Meditation should be without any expectation. Oneness with almighty be from deep inside. It must be total surrender mentally and not superficial physically only.

Let us first discuss stanza 26th of 9th chapter before other stanzas.

Patram pushpam phalam toyam yo me bhaktya prayachchhati I
tadham bhaktyuphritshnami prayatatmnah II26 II

(Gita 9, 26).

This stanza informs that super soul accepts whatever is offered with devotion. This is as per the human tendency, and mind-set which is satisfied with material offering or gift or presentable item. This psychology is prevalent universally. With such a mindset the 26th stanza explains that super soul (parmatma) accepts any offering offered with faith, devotion, and transcendental consciousness. It is a sharing whatever one likes to offer to god. In fact, these are deceptive actions. The give and take actions of devotee, truly is befooling, oneself. But one must be honest in whatever one does. The intensions must be pious in whatever the method he chooses. All austerities performed be dedicated to the Almighty. Thus, one will be free from the results of actions (Karmphal) and through renunciation one will be liberated. Almighty is free from partiality. HE accepts everything offered piously with devout mind set. This is not a freedom from behaving antisocial or anti living beings – creations of super soul (Stanzas 27-32, chapter 9). In the same context the 13th stanza continuous chanting with desire of oneness with

Almighty, is a practice in SANATAN DHARM or eternal basic religion. Hence 14th stanza of 9th chapter of Gita advocates NAMAN to super soul and HIS fractions in all beings:

Satatam kirtayanto ma yatantashchya dridhavratah I
namastanshcha ma bhaktya nityayukta upasate II 14 II

(Gita 9, 14).

NAMAN/Namaste. (highlighted in above stanza), is not physical or showing only but salutation with complete surrender mentally (intelligence) and from heart (emotionally). Hence NAMASKAR is highly scientific, hygienical and maintaining social distance, as per corona pandemic. Without bias there is no ill effect of Namaste greetings adopted by all the humanity of universe. It does not belong to any sectorial religion. After knowing, just a fraction of VED and Puran, Gita author feels Hindu religion is not a religion of Bharatvarsha (India). This word is forced on India by invaders to bring "Bhartiya Dharm – Sanatan Dharm, (not religion) at par with Christian, Muslim, Bouddha, Jain etc. Meanings of words changed to suit the intentions of invaders to make the residents of Bharat country feel inferior. A few words have already been discussed in previous chapters at relevant places, and the vedik meaning emphasized for example:

YADNYA	=	sacrificial offers to fire God – wastage of food into fire.
VARN System	=	cast system to divide population.
ASHRAM system	=	BRHMCHARYA, GRAHSTHA, SANYAS, and VANPRASTH ASHRAM as useless and unnecessary.

True meaning of these words, as per Sanskrit literature be taken into cognizance which has been discussed earlier.

CHAPTER TEN

VIBHUTI YOG
KNOWING ABSOLUTE TRUTH
MANIFESTATION OF NATURE
(BRAHMOTPATTI)
AND LIVING BEINGS
(PRAJOTPATTI)

Abstract: A great, powerful, and majestic person is VIBHUTI in Sanskrit. Lord Krishn is such person in Gita. At many places the words, Me, Mine, My devotee, are used for supreme Soul in Gita stanzas. Gita is knowledge of VED made easy for the general population. Intricacies of stanzas of VED make them confusing and difficult to understand. It is still more difficult to understand the hidden science in the Sanskrit scripts. Even the hidden science in Gita descriptions is difficult to explore. One must ponder on each stanza to find scientific facts in it. Knowledge about supreme soul in its perfection is still unknown to even highly evolved persons of this material world (Gita 10, 2). All GODS (elements) of universe – as clarified earlier (surya devta, vayu devta, agni devta etc.) Even the great sages and RISHI-MUNIS etc. do not know the origin of supreme soul who is the cause of all causes – Gita 10, 1 and 2). By mental speculation one cannot understand HIM. Super soul cannot be conceived by mental speculation. One can only understand HIM in transcendental confidential intelligence in transcendental position, faith, and devotion. Super soul is unborn, beginning less and endless. HE is unborn, beginning less but responsible for creation of the whole universe. Stanzas 4 and 5 indicates that in every stage of evolution HIS presence becomes essential for existence. Hence, HE is omniscient (knowing everything), omnipresent (present everywhere) and omnipotent (skill in everything).

How the population manifested has been mentioned in this chapter. Gita states:

Maharshayah sapta purve chatvaro manavastatha I
madbhava manasa jata yesham lok imah prajah
(GITA 10, 6).

The seven great sages and before that four great sages – MANUs – progenitors of mankind, all are born from the supreme soul.

One in thousand men only achieve perfection. Above stanza (underlined word) speaks that: all living beings are developed "**from my mind**" (**Madbhava manasa jata**). This sounds unscientific. From Gita stanza ch. 10, Stanza 6, author investigated Sanskrit literature elsewhere. It was written with scientific details in UPANISHAD. Author has tried to explain this statement that – "population developed from the MIND of Super soul (Lord Shri Krishna)", from TAITTARIYOPNISHAD (see explanation later in this chapter). After understanding the facts of manifestation of population in Upnishad the MIND factor will be discussed. The discussion is long hence have patience to understand it.

In SHIKSHVALLI of upnishad ANUWAK 3, stanza 1 and 2, this knowledge is in five different groups:

Saha nou yashah I saha nou brahmavarchasam II I II

Adhilokam adhijyotishimdhividyamdhiprajamdhyatmam I
ta mahasanhita ityachkshate II 2 II

The knowledge is explained in 5 different sections:

1. ADHILOK – about worlds or universe.
2. ADHIJYOTISH – about PRAKAASH, Light or Tej.
3. ADHIVIDYAM – about Knowledge – VIDYA.
4. ADHIPRAJAM – about Population – PRAJA.
5. ADHYATM – about self Soul – AATMA.

All this makes MAHASAMHITA. Abstracts are selected and mentioned below.

Athadhilokam I
pruthivi purva lokam I
dyouruttarrupam I
aakashah sandhihi I
vayuh sandhanam I
ityadhilokam II 3 II

About Abode: Dwelling place is Earth (PRITHVI LOK). Heaven (DYU LOK-SWARG lok) is away. But both are joined by Space (Aakash). Air relates or collimates both. As per science atmosphere is 1000 Km. About 15 Km is air. About 200 Km of atmosphere is light or Prakaash. Rest is Dark. This lighted part is Heaven – Swarg.

Athdhijyoutisham I
agnihi purvarupam I
aaditya uttarrupam I
aapah sandhihi I
veidyut sandhanam I
ityadhijyoutisham II 4 II

About Light – Prakash – Tej (Jyoti): Fire is at dwelling place. Sun is away from earth. Water joins both. Electricity relates or collimates both. In stanza above, the word used is Jyoti = TEJ which includes heat and light both. Fire gives heat band light both is at dwelling site. Vegetation or wood collects energy of sun with the help of water.

Without water vegetation cannot store sun energy in wood. The solar energy which is fire in wood. Therefore, water is a joining element. Electricity collimates both means that electricity is responsible for fire on earth. Electricity fell on earth which ignited first fire in dried woods on earth. Sun is responsible for clouds and clouds are from steam (steam from water), clouds are responsible for electricity, hence electricity is due to sun. The electricity collimates fire and sun both. What a scientific derivation?!!!

These are the enigmatic versions of the vedik stanzas in Sanskrit language. It is not easy to understand the hidden meaning in the language used. Probably this may be the reason for thinking as orthodox and useless. The present generation is lucky that such literature was not destroyed by invaders. This foreshadows another warning that countries defences must always be strong for the protection of wealth of knowledge of the nation. How then the RISHIS and Sages be considered as unscientific. One may not understand meaning of their coded versions that does not grant you freedom to consider them as unscientific.

Athadividyam I
aacharyah purvrupam I
antevastyuttarahadobhas rupam I
vidya sandhihi I
pravachanam sandhanam I
ityadhividyam II 5 II

About knowledge: Teacher is at dwelling site. Pupil takes time to learn is away. Subjects join both. Lectures relate or collimate the two.

Athadiprajam I
mata purvarupam I
pitottararupam I
praja sandhihi I
prajanana sandhanam I
ityadhiprajam II 6 II

About living beings: Mother is near, Father is at distance even not known also. Progeny joins two. Reproduction relates or collimates the two.

Athadhyatmam I
adhara hanuh purvarupam I
Uttara hanuruttararupam I
vak sandhihi I
jivhah sandhanam I
ityadhatmam II 7 II

Lower jaw is near because under control. Upper jaw is away, not in control. Voice connects both. Tongue regulates or collimates both.

From science point of view all 5 are important where chronology of birth of universe, light, knowledge, population, and supreme soul has been described. Super self-consciousness only can guide one to understand super self – soul. It has been discussed in 7th chapter previously. Senses have been provided to living beings for survival in material world. Supreme self is far beyond the senses. No sense is perfect to understand super self. Stanza 10, 6 is re-written for ease:

Maharshayah sapta purve chatvaro manavastatha I
madbhava manasa jata yesham lok imah prajah
(GITA 10, 6).

There are numerous scientific facts hidden in this stanza. Author has verified the meaning of this stanza from other sources. What is Sapta rishis? What is MANU? What is abode? What is PRAJA or living beings? How living beings are produced as per the understanding of Sanskrit literature. Detailed search carried out in available literature. Following scientific description is as per available literature:

Seven RISHIS (Sapta Rishi) and four other great sages, and all MANUs (source of mankind total 14 MANU) are super soul's creations (madbhava manasa jata). The living beings, of all the abodes (LOK) of the universe populating the planets in fact descend from them – abodes – fourteen in numbers – (Chouda bhuvan). Details are beyond this book. In 6th stanza origin of population from BRAHMA on wards has been described. Details are available elsewhere. For understanding 6th stanza, one must refer to details narrated in PADMA PURAN and SHIV PURAN. Fraction of energy from super soul which is in the centre of whole universe as SHIV LING composed of (TEJAHPUNJA) fire having heat and light of (KOTI SURYA)- CRORS OF suns, radiation, etc. As in star sun, there are multiple explosions like nuclear blasts emitting fire, heat, radiation, and light. Similar blasts are taking place in self manifest SHIV LING. This SHIV LING is surrounded by primordial gases where VISHNU is lying on KSHIR SAGAR (collection of gases with whitish hue). From the centre of this (UMBILICUS OF VISHNU) energy is emitted as BRAHMA (from SHIV PURAN, 19th Chapter, stanza 1 to 44). Thus, BRAHMA was born from energy of emitting fire, heat, radiation, and light – super soul (HIRANYAGARBH). Seven great sages and 4 other sages (celestial beings) were born from BRHMA and a MANU was manifested. ENTIRE population of universe was manifested from these. Hence these are the patriarchs of living beings. In other word, energy of super soul is responsible for life and all living beings in the manifest universe. It will be of interest here, to mention the 'seven star' cosmic formation in the sky called as "SAPTA RISHI" (Ursa Major see item number 6, 13 below). This 'Saptarishi Mandal' as cosmic constellation is of multiple importance. The origin and derivation of SWASTIK SYMBOL is from the four positions during the whole year of this cosmic star presence, in the sky. (see figures below).

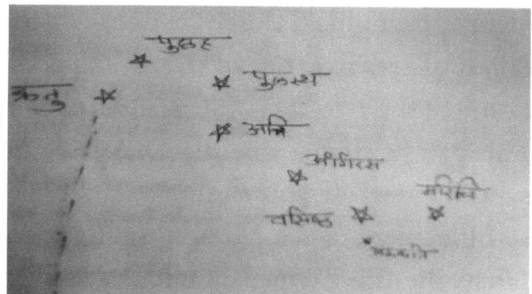

to Polar star

Mizar and its fainter companion star Alcor are easy to spot in the Big Dipper's handle (plough – Ved as SAPTARISHI).

Located in the handle of the Big Dipper, Mizar (brighter) and Alcor (fainter) are one of the most famous visual double stars in the sky.
(Image via ESO Online Digitized Sky Survey.)

Mizar and Alcor appear linked closely, in our sky's dome that they are often said to be a test of eyesight. But in fact, even people with less than perfect eyesight can see the two stars, especially if they are looking in a dark clear sky. This pair of stars in the Big Dipper's handle is famously called "the horse and rider." If you cannot see fainter Alcor with the unaided eye, use binoculars to see Mizar's nearby companion. At the time of year, March, the Big Dipper or URSA MAJOR (saptarishi) is in the northeast. The famous star Mizar is second to the end of the Dipper's handle. Look closely and see Alcor right next to Mizar.
(REFERENCE: Earth Sky: earthsky.org)

VEDIK Knowledge: Alcor ARUNDHATI and Mizar is Rishi VASHISHTHA. Arundhati is faithful wife of Sage Vashishtha. Scientifically speaking these stars are rotating around each other, no one is in the centre or at the periphery, Both rotate around each other. That is why the scientists of Vedik period named these two stars as Vashishtha and Arundhati. Practical use of this in VEDIK marriage ritual is, after marriage the couple is asked to see Ursa Major-Saptarishi specially VASHISHTHA and Arundhati, because both are made for each other and no one is in the centre or periphery in the married life, both are important. Even today this is the practice in VEDIK rituals. The couple of Sage Vashishth and Arundhati possessing high degree of love goodness and exemplar and epitome of married life.

Saptarishi in different seasons of the year Spring, winter, fall and summer seasons. Imaginary diagram to show the saptarishi constellation together seen in different seasons. When joined in the centre by lines as shown in figure above, it takes the shape of a symbol called SWASTIK.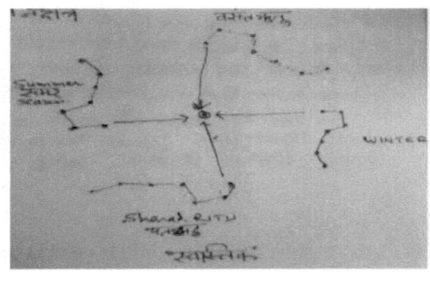

Constellation of Saptarishi in clear sky with names of stars as RISHIS. This symbol is adopted by many countries.

The earth is between North star or DHRUV TARA in north and Agasthya star in south. These are fixed stars and constantly seen at one place in the sky. Sapta rishi constellation has 7 stars. These stars are named after eminent research scientists of Vedik civilization (see figure above).

For interested reader the following description is presented here from SHIV PURAN:

Apart from earth there are more dwelling places in universe according to SHIV PURAN. Distance has been mapped in shiv PURAN. The distance measures and different dwellings are as follows:

Sanatkumar to Sage Vyas explains about Sun, Earth, Planets and stars etc along with 7 mahalok and spread of BRAHMAND: (quoted from Shiv Puran in Italics English below)

1. *The extent to which the rays of the sun and the moon **shine and illuminate** is the "BHULOK" the earth. It is called **Bhūloka.***

2-3. *The sun's sphere is situated a hundred thousand Yojans from the earth. The moon is situated thousands of Yojans from this earth and a hundred thousand Yojans from the sun.*

4. *The spheres of the planets are entirely situated above the moon along with the stars about ten thousand Yojans one above the other.*

5. *Budha (Mercury); above that is Kāvya (Venus): above that is the sphere of Bhauma (Mars). Bṛihaspati (Jupiter) is above that and Śanaiśchar (Saturn) is above that.*

6. *The sphere of the seven sages (The Ursa Major) is situated a hundred thousand Yojans above it. The Pole Star (or Dhruva) is situated a hundred thousand Yojans above the sages.*

7. *The Pole star is a meritorious star and in the center of the circle of luminary bodies (Center of Jyitishchakra). The **Bhuvarloka and Svarloka** are above the earth but beneath the Pole star.*

8. *A crore Yojanas above the Pole star is the **Maharloka** where the seven sons of Brahmā are stationed. They live even during the time of Kalpa.*

9. *The seven sons are—Sanaka, Sananda, Sanātana, Kapila, Āsuri, Voḍhu and Pañcaśikha.*

10. *Above that, is stationed Śukra (Venus) at a distance of two hundred thousand Yojanas. Two hundred thousand Yojanas below it is stationed Budha (Mercury) the son of the moon.*

11.   O sage, Bhauma (Mars) is stationed two hundred thousand Yojans above it. Bṛhaspati (Jupiter) is stationed two hundred thousand Yojans above it.

12.   Two hundred thousand Yojanas above Jupiter, Saturn is stationed. These are the seven planets that stay in their respective zodiacs.

13.   The seven sages (Ursa Major) are stationed eleven hundred thousand Yojans above the planet Saturn. The Dhruva star is said to be stationed fourteen hundred thousand Yojans above the great star.

14.   Sixty-three hundred thousand Yojans from **Janaloka** is stationed the Tapoloka where the gods called Vairājasstay. They are free from burning sensation.

15-16. Eighty four thousand Yojans from **Tapoloka** is stationed the **Satyaloka.** It is known as Brahmaloka where the pure-minded, perfectly wise Brahmachārins engaged in truthful virtue stay. Men who go there from the earth also stay there.

17-18. In the Bhuvarloka, the sages of great Siddhis stay in the form of gods. In the Svargaloka, the gods, the Ādityas, the Maruts, the Vasus, the twin Aśvins, Viśvedevas, Rudras, Sādhyas, Nāgas, Khgas etc. stay. Above that are the nine planets and above them are the seven sages free from sins.

19.   O sage, thus the seven worlds have been mentioned to you. The Pātālas too are also seven. Thus the universe has been explained to you.

20.   As the wood-apple is both above and below the earth, so the universe is enveloped by the cauldron of the Cosmic Egg.

21-23. It is surrounded by water ten times as large all round the fire, wind, ether and darkness. It is encompassed by the elements and the principle of **Mahat**, eight times as large. There surrounded by the Mahat and Pradhāna is stationed the Puruṣa. **The supreme soul is infinite. He is therefore called eternal since he has no limitation.**

24-27. This eternal is the cause of all. O sage, it is the great Prakṛti. **From that infinite of unmanifest origin thousands and hundreds of thousands of cosmic eggs are born.** The self-known great soul contains everything just as the wood contains latent fire, the gingelly (Sesame) seed the latent oil and the milk the latent ghee. From this primordial seed emerge all those beginning with Mahat and ending with Viśeṣa. Then the gods and others are born. Then are born birds, their progenies and the seeds of others.

28-29. When the tree comes out of seed it docs not suffer any depreciation or diminution. Just as fire appears from the solar stone coming in contact of the sun so also the creation takes place. Śiva has no desire for it. As a result of the contact of Śiva with Śakti, the gods and others are born naturally.

30.   They spring up through their own action. Śiva manifests himself as Brahmā, Viṣṇu and Rudra.

31.   Every thing springs up from him and finally merges in him. Śiva is sung as the performer of all activities.

Sanat kumar further says to Vyas that there are more lok beyond that too.

Different (lok) mentioned above in UMA SAMHITA are summarized below:-

The names of these 14 LOK or abodes are: below BHU LOK. These are: Atal, Vital, sutal, talatal, mahatal, rasatal and pataal. Above BHULOK, the abodes are named as: BHU, BHUVA, SWAH, MAH, JAN, TAP, and SATYA LOK. Location of these abodes are mentioned below (the scale is as per norms of the vedik period):

One lac yojan away from earth (BHU LOK) in a circle of 1000 yojan, is placed "solar system". At 1,00,000, yojan distance is placed SAPTARISHI MANDAL and from here 100,000 yojan away is placed POLAR STAR (DHRUV TARA). This is in the centre of galaxy. Beyond POLAR STAR is one cror yojan away is MAHAR LOK. Beyond this at 2,00,000 yojan Venus, at 2,00,000 yojan below Mercury (born from Moon), 2,00,000 yojan above Mars, 2,00,000 yojan away is Jupiter, 2,00,000 yojan is Saturn all in their orbits, which is JAN LOK. From here 4 and1/2 times away is TAP LOK, where VEIRAJ (Devta) is situated who is without heat. From TAP lok, about 6 times away is SATYA LOK, is situated. All this is endless and cause of all universe and nature. Beyond this also there are dwellings, called as VAIKUNTH, KAUMAR lok, UMA lok, and SHIV lok

(This description exists in SHRI SHIVMAHAPURANANK UTTARARDH, Gita press Gorakhpur. No. 1, Year 92, Page 302, 19ᵗʰ chapter, 5 Uma Samhita, of Shiv puran.)

Sanskrit speaking civilization's distance Measurement Chart for calculations given below (for understanding YOJAN measurement):-
(The "Artha shastra: Chapter XX. Measurement of space and time", authored in 4ᵗʰ century BCE by Chanakya (Vishnugupta Kauṭilya), sets this standard breakup of Indian units of length:)

⅄ 1 Angul (approximate width of a finger) = approx. 3/4ᵗʰ of an inch;

⅄ 4 Angul = Dhanurgrah (bow grip) = 3 in;

⅄ 8 Angul = 1 Dhanurmushti (fist with thumb raised) = 6 in;

⅄ 12 Angul = 1 Vitastaa (span-distance of stretched out palm between the tips of a person's thumb and the little finger) = 9 in;

⌖ 2 Vitastaa (from the tip of the elbow to the tip of the middle finger) = 1 Aratni or Hast (cubit or Haath) = 18 in;

⌖ 4 Aratni (Haath) = 1 Dand or Dhanush (bow) = 6 ft;

⌖ 10 Dand = 1 Rajju (also called *Hasta* or *Haath* हाथ, i.e. hand) = 60 ft;

⌖ 2 Rajju = 1 Paridesh = 120 ft;

⌖ 2000 Dand (Dhanush) = 1 Krosha (krosh or kos) or Goruta = 4000 yards or $2\frac{1}{4}$ miles – nearly 3.66 km; *4 Krosh = 1 Yojan = 8 miles – nearly 13 km;*

Apart from performing duty assigned as a human being the statements in GITA are for liberation of human race and individual man. Constant transcendental involvement in super soul frees persons from bad effects of deeds of that individual. **Physical body is important for liberation. Material body is a vehicle to carry out aims in life. Without body transcenden tal meditation and understanding of super soul is impossible. Therefore, healthy body is essential. For spiritual progress to pursue, healthy body is first and most important means.**

Before this one must understand the formation of body. How the material physical body or KSHETRA media for appreciation of soul and super soul? Gita stanza 8 of 10th chapter says:

Aham sarvasya prabhavo mttah sarvam pravartate I
itimatva bhajantemam budha bhavsamanvitah II 8 II

(Gita 10, 8).

The spiritual and physical manifestation in this universe is due to the supreme soul. It functions due to super soul. Intelligent and knowledgeable persons understand this and get attached to super soul. The scientific hidden meaning is explained as under:

FORMATION O F BODY: It is interesting to note here that the rishis or scientists of Sanskrit speaking civilization knowing fully about sperm (SHUKRA/) and female (RAKTA BEEJ) element in the formation of body of living beings which is from material world. The knowledge of physical or manifest form in the material world is getting confused du e camouflaging effect of the whole universe (MAYA). Modern science understands only presence of Sperm and Ovum. The Sages and Rishis-Munis also investigated that how these basic sperm and ovum are formed. How the material elements came to exist in material universe? PRANIK ENERGY (positively charged) and RAYI (negatively charged) coming together to form body. The creation and effectively getting involved in the automation effect of the creator of universe or BRAHM. Hence apart from physical existence the spiritual existence also deeply studied. Fortunately, modern scientists have started believing this aspect of existence of our body in material world.

This raises one question that is evolution from one source and progressing to human being becomes questionable – Monkeys to man. There are different YONIS and different form and different shapes are created through different material nature (Gita 14, 3-4). This sounds different opinion in comparison to modern theory of evolution. Of course, formation of material body from different yonis and progressive evolution is the prerogative of all formations.

Mam yonirmahadbrahm tasmin garbha dadhamyaham I
sambhavah sarvabhutanam tato bhavati bharat II 3 II

(Gita 14, 3).

Material NATURE is the main source where super soul sows seed and thus from union of material and CHETAN-energy, living beings are generated. Details given below.

Sarva yonishu kounteyo murtayah sambhavantiyah I
tasam brahmamad yonirham beejpradah pita gi II 4 II

(Gita 14, 4).

In all living lineages bodies are formed. Different forms and shapes of bodies are created through different material nature. Super soul being father in the form of seed.

A clear understanding of body formation can be grouped into:

1. **Pre embryologic formation.**
2. **Embryologic formation. (Development of fetus in Uterus)**
3. **Post embryologic formation. (After delivery also the growth continues till full grown animal/man)**

Following expression is just a small summery of the subject:

For this description author has taken help from AITAREYOPNISHAD.

Pre-embryologic formation includes – creation of nature – BRAHMOTPATTI, and Creation of living beings PRAJOTPATTI. Both creations are inter related.

Creation of Nature or PRAKRITI is explained in RIGVED mandal 10 sukta 190-1, 2, and 3,

Rutam cha satyam chabhidhattpasodhya jayate I
tatoratrya jayat tatah samudro arnavah II 1 II

Ritam is truth which is name of universal laws of nature. The creation of universe is the result of existing source of energy and knowledge of supreme soul or paramatma (performance of TAP by VIDHATA). In other word Self manifest, parmatma is responsible for the creation of SRISHTI, nature, natural laws, universe, day and night and Oceans etc. etc.

Samudravarnavadadhi samvatsaro ajayat I
Ahoratrani viddhadvisvasya mishato vashi II 2 II

After creation of oceans living beings and time established.

Without creation of NATURE (PRAKRITI), living beings cannot be created. Everything in this material world is created with effect and change in basic elements. This depends on:

1. basic VIRTUES (Dharm parinam) effect,
2. effect of characteristic symptoms (lakshan parinam), and

3.   effect of stage or phase (Avastha Parinam) of action and reactions of elements.

This is applicable in all the cosmic and material universe. In Sankhya and yog Sidhant, whatever is generated, causative factor exists and after its end it (causative factor) still exists.

Eten bhutendriyeshu dharmalakshanavasthaparinama
vyakhyata II13II

Yog Darshan, Vibhutipad 3,

(Yog Darshan by Patanjali by Harikrishnadas Goyandka. Code no.135. Gita Press Gorakhpur Vibhutipad sutra 13. Page 489).

Interaction of virtues (Dharm Parinam) is when virtues (dharm) of one element disappears into virtues of other interacting element. In other word, the virtues merge and suppress into virtue of other in creating new element. Practical example can be cited here is of MILK. Dharm of milk is to maintain it, in the form of milk. As per the particle physics, the electromagnetic force maintains the milk state as far as possible. But when curd is mixed it converts into Curd. The milk virtues transform into virtues of curd-a new element. Time is needed to convert milk in curd. (Time factor is important). This is applicable, uniformly in all the elements of the universe. Similarly, characteristics also change with Virtue effect (Dharm parinam). So is the state or phase (Avastha parinam). At the same time, the element never disappears but get transformed. Past and future properties are also present in present form but unmanifest and are subject to change.

Atatinagatam swarupto styadhvabhedadharmanam II 12 II

(Chapter 4 - Kaivallyapad).

Past and future virtues remain in present form and change in dharm, it is dependent upon time (or Kaal). Time makes the difference (highlighted part of stanza above)

(Yog Darshan by Patanjali by Harikrishnadas Goyandka. Code no. 135. Gita Press Gorakhpur Kaivalyapad page 568.)

All this description is as per the modern science and principles.

PRAJOTPATTI (Manifestation of all Living beings):

As indicated in earlier chapters that, Shiv ling in form of fire AGNI SHIV LING which is manifest, having heat and fire. There are gases around it. These are PRIMORDIAL gases. Modern science accepts it. These gases are in the space-Aakash. Both these i.e. Aakash and primordial gas created by super soul. It is called as AMBH in the upnishad.

Primordial gasses + Aakash = AMBH. Movements in AMBH resulted into gravitational force due to which heavier things moved to center and lighter to periphery. This resulted into ball formation. Pressure of gases resulted into gravitational force and heat. This TEJ is named as MAREECHI. This resulted into elemental inter-reaction. These balls are having heat and light-Tej named as MAREECHI, as above or ANTARIKSH.

Inside Ambh gaseous pressure and gravitational force = TEJ. Ambh – pressure of gasses + Gravitational force = Mareechi (Tej).

Movements inside = Ball of gasses. Elemental reaction inside = Balls of heat and light =Antariksha.

After AMBH, and ANTARIKSH, MORTAL WORLD or MAR LOK was created i.e. EARTH like planets or PRITHVI.

Sequence is AMBH > TEJOMAY balls or SUN like balls > Earth like planets which is accepted by modern science. Some interaction on earth planet created Water element JAL TATTVA. This is called as AAP. **Ambh (Gas+Akash) + Mareechi (Tej) + Antariksha = Balls (Sun) = Mar Lok = Planets (Earth) = Water.**

From Aap super soul created energy force called Chaitanya. In AITARIYOPNISHAD 1. 1. 2. Stanza is like this, Stanza by Aittar RISHI says:

Sa imamllokan srujat I
ambho mariachi-rmar-mapo dombhah paren divam dyouh
pratishthantriksham marichayah I

pruthivi maro ya adhastatta aapah II
(Aeitareya Upanishad chapter 1, Khand 1, Anuvak 2)

In next stanza it is clearly indicated that for incubation or ripening or maturation of egg for bird formation, bird sits on egg providing heat, similarly Chaitanya was created by providing heat to Aap. The soul by providing heat ripened Aap: or incubated Aap or water ripened up by Chaitanya in a good manner i.e. to incubate in a good manner needs time. In due course of time mouth appeared (mukham nirbhidyat) like Egg (yathandam). Through this mouth, bird hatches out.

Tambhyatpat I
tasyabhitaptasya I
mukham nirabhidyat I
yathandam I
AITARIYOPNISHAD chapter 1, Khand 1, Anuvak 4.

This sequence is accepted by modern science. A scientific question comes to mind is how this was conceived by Rishi of Sanskrit speaking civilization. All this needs targeted research. May be the literature burned by invaders had explanation in original hand written manuscripts.

In stanzas which follow, the example of formation of man (human embryological development) the formation of Brahma and Prakriti has been compared and explained. For details read the original work in upnishad. Formation of man and manifest world (prakriti) have similarities in the formation.

First part (highlighted-in bold letters) of this stanza is foetal development and the next about manifestation of Nature.

Mukhad vak I vachah agnihi I
Nasike nirbhedyat I nasikabhyam pranah I pranad vayuh I
Akshini nirbhidyat I aksibhyam chakshuh I chakshush aaditya I
Karnou nirbhidyat I karnabhyam shrotram I shrotradidashah I
Tvak nirbhidyat I Tvacho lomani I lomabhya
aoushadhivanahspatayah I

Hrudayah nirbhidyat I hrudayat manah I manash chandrama I
Nabhihi nirbhidyat I nabhyah apanayat I apanat mrutyuh I
Shishna nirbhidyat I shishnat retah I retasah aapah I
(AITARIYA UPNISHAD chapter I, khand I, anuvak 4).

The order of development of tissues and organs of embryo, in embryology and its sequence is taken as example and manifestation of different material universe is compared. The sequence of development is as per modern embryological knowledge. First mouth appears during development which can produce sound, which are sound waves, similarly in developing nature/PRAKRITI sound manifests, which with electromagnetic waves of sound can produce fire, and then manifest fire.

Development of nostrils in embryo similarly in nature manifests PRAN and then wind containing Oxygen manifests. This indicates that the oxygen is manifested later. Modern science accepts this (see reference below).

In Chemistry section of OYLA MAGAZINE 15 March 2020 issue, page 14, (OYLA Ias was the first gas ndia.org), it is clearly mentioned that the first place in atmosphere is N_2, Nitrogen gas. (details about Pushna Vayu in Pushna Chakra).

Eyes generated in embryo. Vision needs light to see so the nature generates Sun.

Ears generated in embryo which is for listening and balancing (semicircular canals). **This gives positioning appreciation and direction to body. Similarly, nature manifests direction** providing DISHA i.e. north, east, west, south, and so on. Rishis were knowing that the ear is responsible for balancing and maintaining equilibrium, direction (disha). What an advanced scientific knowledge!!

Skin develops in embryo with hair on skin. Similarly, in nature all vegetation manifests.

Development of heart and later developed mind. Mind of nature develop Moon. Mind is related to brain. How, it is connected with the heart? This needs explanation. As mentioned earlier in detail that the word heart in SANSKRIT is HRIDAY (हृदय) is as per its function. HRI (हृ) means to receive, DA (द) means to 'give' and y (य) means to go/circulate. i.e. **whereever, there are these three functions mind must be there to control such functions.** Here AITAREYA rishi says SUN manifests from eyes of nature and moon from mind. Rishi mentions here that sun and moon are not from same source. But modern scientific knowledge is all planets of solar system have originated from sun. That is moon is part of sun. But ROCKS brought from moon's surface, during lunar mission of NASA, USA., are different in composition when analysed. This needs more research. But it looks sun and moon are not from the same source. (As per vedik knowledge the moon is from NAKSHATRA – it is explained later).

Umbilical development in embryo in 2nd month, later it connects with mother's circulation and nourishment depends upon blood of mother. **Before that primordial stem cells have capacity to divde and multiply and generate their nourishment to form tissues, is known. These cells have no death.** The cells lateron, for nourishment, depend upon placental blood, this means if no blood from mother death takes place. That is death manifests at this juncture. The death manifests when umbilicus develops. One name of PRAN is APAN which flows downwards (see above). This APAN helps digestion i.e. it destroys food ingested to separate essential elements needed for body. Similar action is likely to be on body as well. In other word, it can be the cause of death to body. This means after APAN death can manifest. It can be concluded that the RISHI's statement sounds scientific and correct as per modern science, but it needs more research. YOG-Shastra confirms this fact. How? *According to YOG Shastra One can defy death by controlling the APAN VAYU. This APAN needs to be reversed. For details see YOG Shastra.*

After umbilicus genitals (SHISHNA) appear in developing embryo. This is in third month of pregnancy i.e. after the development of umbilicus in 2nd month. Genitals are for generating progeny or creation. Similarly, the nature or PRAKRITI manifested reason for creation i.e. water or AAP or JAL with liquid property. All living beings born out of water is known knowledge in modern science.

From modern embryological knowledge it is clear, that the whole process of development of embryo and fetus is based on automation process. Some of the automation factors have been investigated and established to some extent. These are inducers and organizers. The function needs to be regulated. It is by regulation of induction (regulators). At the same time surrounding tissues and cells also have effect on development of fetus. In absence of such factors the fetus develops abnormally. More or details is beyond the jurisdiction of this book (refer to appropriate books). Fetal development in Sanskrit literature is mentioned later in this chapter.
(Human Embryology by Inder Singh, 5th edition. Ch. 21, Some General Considerations. Macmillan India Ltd. 1991.)

In the context of knowledge written by Rishi Aitareya, comparison of development of fetus and the Nature is only possible when knowledge of embryology is fully known. **This means embryology was well developed before Rishi Aitareya. Factors responsible for automation process involved in development of nature, must be there.** *In nature nothing happens without reason. Therefore, these reasons can be investigated and researched out.*

Broad categories of body are:

1.   Physical
2.   Spiritual

Physical body is from food. Food consumed becomes Juice and Residue (किट्ट). Manifestation of body has been studied by modern science, in intra-uterine development growth of fetus or embryonic growth. But formation of body studied by Sanskrit literature even before the formation of embryo in uterus. The study carried out in two phases, pre-embryonic phase and embryonic phase.

PRE-EMBRYONIC PHASE: JUICE: This flows in Channels of body. It is distributed to all body parts. PRAN VAYU + PRAN through Blood. What is the knowledge about PRAN (PRAN DNYAN)? It is essential for persons interested in YOG and AASAN. PRAN is not respiration. It is the root element, generated in the beginning of material world in universe. All 5 elements and every part of material world are generated from space with the help of PRANIK energy. Living beings survive with this PRANIK energy. At the time of annihilation due to non-availability and no support from this PRANIK energy get back into space (– AAKASH). All of material world get manifest from SPACE and disappear in space (CHHANDOGYOPNISHAD):

Aakash asya lokasya ka gatirityakash itihi:

Sarvani ha va Imani bhutanyakasadev samutpdyante I aakasham prastyatam yantyakasho hyeveibhyo jyayanakashah prayanam

(chhandogyopnishad I. 9. I).

All beings of material world, (extensive, presence is considered in its maximum), is in space or AKASH and maximum energy is PRANIK energy. Life of all beings is PRAN. With help of PRANIK energy living beings can breathe. In other word, in all material world pranik energy is existing. It needs clarification that Aakash is divided in two – Space entrapped in body (bhutakash) and the other is free space which includes or support the whole universe which is implicit or indefinite (vivakshit). In stanza 1 of 1. 9 of Chhandogyopnishad (above), it is the second one has been referred. It clarified in Chh. Up. 8. 14. 1:

Aakasho vein am namrupayornirvahita te yadantaratdbrahm tadamrut >\sa aatma prajapateh

(Chhandogya. Upnishad. 8. 14. 1).

All beings i.e. basic elements, in material world manifest from PRANIK energy in Aakash and un-manifest in same energy. During time of manifestation, (day of life of Brahma) everything manifests from Pran and during annihilation time (Night of Brahma) all unmanifest into Pran. Therefore, the "Pran Devta" is followed in the proposal (**stanza below**).

Pran iti hovach sarvani ha va Imani bhutani
pranmevabhisamvishanti I
pranamabhyu jjihate I
seisha devta prastavmanvayatta I
tam chedvidva nprastoshyo murdha te
vyptishyattthoktasya mayati I
(Chhandogyopnishad 1, 11. 5).

Prano brahyeti vyajanat I
pranadhyeva khalvimani bhutanijayante I
pranen jatani jeevanti I
pranam prayantyabhisamvishntiti I

In PRASHNOPNISHD, while answering the question that from where the living beings are generated. It is expressed and described regarding manifestation of nature that is creation of RAYI and PRAN. "RAYI" (Matter) is negatively charged and PRAN (energy) is positively charged and union of two living beings are manifested in this world:

Tasmei hovach prajakamo vei prajapatihi sa tapotapyat sa
tapastaptva sa mithunamutpadayate I
sa meithunamutpadayate rayim cha pranam chetyou me bahuddha
prajah karishyat iti I
(PRASHNOPNISHAD Question 1, stanza 4).

Initialy the Prajapati was alone and desired to create Living beings so he tried his resources like energy, knowledge etc. (tapotapyat sa tapastaptva). Prajapati churned austerities of body and mind. He generated heat due to his efforts or attempts. From this RAYI and PRAN pair could be manifested.

What is RAYI? In next Stanza 5, **of Prashnopnishad** the answer is like this:

Aadityo ha vei prano rayirev chandrama rayirva I

etatsarve yanmurtam cha tasmanmurtirev rayihi II 5 II

(Prashnopnishad, 5)

Highlighted part of stanza above, is final version. All manifest or unmanifest (अमूर्त) elements are RAYI (Matter) and have PRAN (energy) in them. **In other word, the rays and energy emitting out of all elements are Rayi. Combination of Rayi and pranik energy are responsible for generation of living beings.** All elements in space (earth, water, Air, Fire, Space, Light, big and small groups of Stars) emit energy and rays, are called RAYI and ENERGY (PRAN) due to which the vibrations, movement is happening, have come to workable stage, is called PRAN. *Manifest and un-manifest substance is RAYI, while PRAN is "life force". Engineering of all this is facilitated by the ENGINEER is called as HIRANYAGARBH or PRAJAPATI.* This explanation sounds perfect and scientific but needs a lot more research.

Scientific details might have been there in original manuscripts in nine story building of libraries of NALANDA and TAKSHSHILA burnt by invaders, the fire continued for three months. Oh! what a loss of knowledge of humanity which was for the good of humans!! This loss is irreplaceable. This loss is due to ignorance and foolish CHITTA VRITTI of invaders under the influence of material greed.

There are five types of CHTTAVRITTI of mind set (explained later). Human beings are not equal in mental development and are at different level of mental state. Starting from brutal animal level to philosophic attitudes. Accordingly, their actions are observed in this world (explained earlier).

EMBRYONIC PHASE: In SHAKTI PURANANK – part of MADBHAGWAT details of intrauterine development of growing embryo has been described. This is in the form of dialogue between daughter and father. Daughter (illumined incarnation of Goddess), explaining to her father about it. Wholesomely and comprehensively PRAN is running the show of all universe. Not only human body but all living bodies are getting life using this PRANIK energy. The work of all senses is effectively carried out by consuming PRANIK energy. Therefore, one must understand the PRAN. Scientifically as per the work carried out by PRAN, has been named accordingly. Various names of Pran their functions and flow channels are described below:

Pranopanah samanoshchodanvyanou cha vayvah I
(PRAN, APAN, SAMAN, UDAN VYAN).

Nagah kurmoth krukaro devdatto dhananjayah II
(NAG, KOORM, KRUKR, DEVDATT, DHNANJAYA) YAGYAVALKYA
(Yogi Yagyavalkya chapter 4, stanza 66-69)

Nihishvasochchhvasakasashch prankarmeti kirttirtah apanvayoh
krmetad vinmutradivisarjanam

Hanopadancheshtadi vyan karmeti cheshyate udankarmatat
proktam dehasyonnayanadi yat

Poshanadi samanasya sharire karma kirtitam udgaradi gunoyastu
nagkarmeti chochyte

Nimilnadi karmasya kshutam vei krukarasya cha devdattasya
viprendra tandri karmeti kirtitam

Dhananjayasya shokadi sarvakarmaprkirtitam

PRAN = It is situated from heart to up to nose. Senses of upper body are dependent on this. Breathing, control of rhythm of entry – inspiration and expiration, water food digestion, water into sweat, urine, semen etc. is work of PRAN-VAYU.

APAN = excretion of urine, stool, semen, moving fetus down from uterus. It is situated from umbilicus to feet. Lower senses are under control of this.

SAMAN = Central part of body from umbilicus to heart. To distribute equally the juice of food to whole body.

VYAN = Its place is from root of genitalia upwards. Its main work, is blood circulation.

UDAN = it is placed from neck to head. It is connected to PRAN of whole universe. This at the time of death PRAN expelled from body and at time of fertilization it gets into uterus. By YOGIK technique PRAN can come out of body and roam around and get back to body without any adverse effect on physical body, and body, continue to remains living, till such time, the PRAN returns back into body.

NAAG = sneezing, coughing, **KOORM** = contraction, Hiccup etc. **KRUKAR** = thirst, hunger etc. **DEVDATTA** = sleep, lethargy, sleepiness, **DHANNAJAY** = nutrition of body even after death.
It is an understanding that if one controls PRAN controls his body and senses of the body. Control regulation of PRAN is called PRANAYAM. The technique of regulation and control of PRAN is described in PATANJAL YOG SHASTRA.

RESIDUE is excreted in 12 forms (Through Eyes, Nose, teeth, tongue, Skin, Nails, Genitalia, Anus and others are sputum, sweat, food residue – night soil and urine

To maintain healthy, clean, pure and body free from disease. Detailed techniques and methods are described elsewhere in PATANJALI YOG SHASTRA and PATANJAL YOG PRADEEP. To summarize in short: Techniques of HATH YOG, Natural Medication and determination are essential. These resources and means are useful only when body is healthy and appropriate to follow the above mentioned, means. Difficulties and disruptions which arise in YOG SADHNA i.e. pursuing YOG techniques are:

Vyadhistyan sanshay pramadalsyavirati bhranti darshnalabdhabhumi katvanvasthitatvani chittvichha shakshepastentarayah
(30, Samadhipad, Patanjal YOG Pradeep).

Some of the disruptions are: Ailment, Doubt, Laxity, Laziness, Addiction, Fallacy, Misunderstanding, Unsteadiness, disappointment when not achieving SAMADHI etc.

Physical body and formation of body: Physical body is formed from food. Essence of food is responsible for the formation of Semen and sperm, and sperm forms body after uniting with ovum. At this union, the composite form of PRAN and fraction of super soul – the soul enters the body part of female. Apart from this another interesting part of this union is – how the parents are chosen? It has been explained. Entry to and selection of particular parents, depend upon his DEEDS – KARM and KARMAPHAL, results of his past actions deeds and work done.

The development of foetus is explained:

Day 1, it becomes KALAL, in five nights it becomes BUDBUD, in seven nights it becomes fleshy-MANSPESHI.. After this formation, neck, head, shoulder, back, spine. Waist etc. in a time span of two months. By 3 months all joints are formed. By 4 months fingers are formed. In 5 months, mouth, nose, ears, dental line, anal region, nails are formed. By 6 months ear hole is formed. By 7 months rectum, penis, umbilicus etc. develop. Thus, fully developed foetus enveloped in membrane. It develops and remains in uterus. After this memory develops.

Interestingly, a fact disclosed here, is that pious human is delivered in 7 months (because of size of foetus delivered easily without any pain or difficulty) and rest undergo torture of delivery. Aim of life is to have long, healthy, disease free life. Control of PRAN i.e. PRANAYAM practice can provide long, healthy, disease free life. Apart from this it can help to get success on control of mind and senses. Control of mind is the most difficult. Mind and PRAN have great relationship. Control of PRAN tame mind. Mind is basically restless. Hence must be controlled for overall growth and welfare. Therefor PRANAYAM is essential for YOG SADHANA.

After understanding different works of different PRAN as mentioned above, it is mandatory to know that how the PRAN flows in body.

Invisible PRAN flows into channels called NADIS in Sanskrit. There are main 15 NADIS in human body. Some of them can be correlated with modern anatomy but others are still to be recognized. Sanskrit names of these channels are:

1. SUSHUMNA.
2. EIDA,
3. PINGLA.
4. GANDHARI.
5. HASTAJIVHA.
6. PUSHA.
7. YSHSVINI.
8. SHURA.
9. KUHU.
10. SARASVATI.
11. VARUNI.
12. ALAMBUSHA.
13. VISHVODARI.
14. SHANKHINI.
15. CHITRA.

Out of these 15, YOG is closely related to 3 main NADIS, namely SUSHUMNA, EIDA, PINGLA. Out of these three SUSHUMNA is the best of all. It starts near anorectal region and flows through spinal column and reaches brain. From this region (anorectal), on left side of SUSHUMNA (– inside spinal column), EIDA (left sympathetic) and on south side of PINGLA (right sympathetic) starts and reaches root of nose. At this point the three channels meet. Interestingly SUSHMA, EIDA, PINGLA also recognized as SARASWATI, GANGA, YAMUNA respectively. (these are also the names of rivers of BHARATVARSH) These channels at the starting point are separate hence named as MUKTA TRIVENI – separate three. At the root of the nose the three channels, meet, hence called as YUKTA THRIVENI – meeting of three Nadis. It is just like three rivers unite at PRAYAG city. Modern name is

ALLAHABAD city. It is called as TRIVENI SANGAM. Similarly, like breathing, where air flows in by inhalation and expelled out by expiration, the PRANIK energy flows through channels called as EIDA and PINGLA NADIS. EIDA and PINGLA are recognized as CHANDRA NADI and SURYA NADI respectively. Sometimes breathing is through left nostril and some time through right nostril. Every person can experience this. When breathing speed is faster through left nostril it is recognized as EIDA or CHANDRA SWAR. When breathing speed is more through right nostril it is called as PINGLA or SURYA SWAR. But sometime the breathing is equal through right and left nostril it is called as SUSHUMNA SWAR. During day and night, breathing through right and left nostril takes place twelve times systematically, serially and respectively. This is in an orderly manner. It is advised and is beneficial that during Joint period of day and night YOGA exercise be carried out when SUSHUMNA SWAR is active. Such explanations are provided in modern medical sciences.

aadou chandrah site pkshe bhaskarastu sitetare I
pratipada dinananyahustrini trini kramodaye II34II
(Patanjal yogpradip sutra 34)

This has been explained in Chapter 4 of GITA (SHLOK 27-31).

Sarvanindriyakarmani prankarmani chapare I
aatmsamyamyogagnou juhvti dnyandipite II27II
(GITA, 4, 27)

Self-realization is achieved by mind control and control of senses.

Dravyayadnyastapoyadnya yogyadnyastathapare I
swadhya yadnyana yadnyashcha yatayah sashitvratah II28II
(GITA, 4, 28).

Enlightenment is possible by sacrificing possessions, by performing severe austerities, by practicing yog, by studying VED to advance transcendental knowledge.

Apane juhyati pranam pranepanam tathapare I
pranaprangati rudhva pranayamprayanah I
apare niyataharah pranan praneshu juhyati

(Gita 4, 29)

Some practice breath restraint, and offer incoming breath in outgoing breath, and remain in trance. Others control diet.

Coming back to stanza 6, of chapter 10, of Gita mentioned in the beginning of this chapter that the living beings in the universe are manifested from the mind of Brahma. What is mind? What is the meaning of mind? Where is this sense organ situated anatomically? To understand mind YOGVASISHTHA is helpful. This is collection of teaching lessons by Sage Vasishtha to disciple RAM. The lessons were delivered for 18 days from sun rise to sun set every day for 18 days.

Jayate man eveh mana eva vivardhate I
samyagdarshandrushtya tumna ev hi muchyate II

(Yog vashishth 5. 11, 14).

It is the mind which takes birth in this world. Grows and swells. Birth, growth and freedom are of Mind and nothing else. Mind is not substantial, it is imaginary. The mind is a process of imagination. As super consciousness is full of self – revelation. Mind has no separate entity. It is a thought issuing from supreme consciousness. Imagination or SANKALP gets to know of consciousness, it becomes the super and pure consciousness:

Man evamsankalpam chitprasaden jeevati I
bhavyan vishvameveikam chintametya chidpyut II 49 II

Dhishchittam jeev ityetah sankalpam syasato matah I
sandnyah sankalpitastajdnyeirna ram parmarthatah II 74 II

(Yogvasishth 13.)

In chapter 14, manifestation of Brahma from navel of Vishnu (primordial gasses), creator and preserver of all from supreme spirit, has been described.

From the mind (as above) of Brahma manifested prime progenitor of mankind. Because manifested from mind is called MANASPUTRA. In the same context Lord Krishna explains in Gita 10. 6, the universal population and its manifestations. From energy of HIRANYAGARBH the supreme lord, Brahma manifested. From Brahma seven rishis and four great sages and MANUS manifested. Total 25 in number. These are the patriarchs of the living beings of the universe. There are innumerable universes and innumerable planets in each universe. Each planet is having populations of different varieties. Origin of these is through these 25 patriarchs. Gita Chapter 11 stanza 39 says that Brahma is grandfather and the supreme soul is great grandfather.

Scientifically all sounds imaginative without any proof. But there are elements, PANCH MAHA BHUT etc. Evidence suggests their existence in material world (Marishi Kanaad's Vasheshik Darshan and Nyay Darshan as in previous chapters). This knowledge may be from transcendental meditation or devise not known to us at present. As we do not have knowledge about birth of 101 children outside human body in a device can be called "artificial uterus".

If one considers the world, it is a display of Nature. If one thinks about world, the distinct understanding comes forth is what one sees is the world and the other is with which one sees the world is energy, life force or the soul or self or AATM. One is external body and the other is invisible, which is internal, one can name it MIND. Body cannot do anything without the MIND which exists within the body. Body functions for the sake of mind and helped and guided, controlled by mind. Body has parts, organs but the Mind is subtle, internal and invisible unlike body. Body and soul are like container and contained. This is what Sage vashishth explains to Lord Shri Ram in YOG VASHISHTHA chapter 71, stanza 60:

Aadharadheyayoreknashe nanyasya nashtata I
yatha tatha shriradinashe natmani nashtata II60 II
(Yogvasishtha Ramayan, 71, 60).

If container is destroyed the contained is not lost (Aadharadheyayoreknashe). It is like if green leaf dries up the water is not lost but evaporated into sun's rays. Similarly, when body is destroyed the soul is not (Stanza 61).

Ekaparnarase kshine rasoneiti yatha kshayam I
yati prnarasashcharkarashmijalantare yatha II6I II
(Yogvasishtha Ramayan, 71, 61).

In short, as the everywhere existing sky is reflected in mirror, similarly omnipresent and omniscient soul is reflected in MIND (Stanza 38 of chapter 71 0f Yogvasishtha Ramayan):

Sarvatra sthitamakashamadarshe pratibimbati I
yatha tathatma sarvatra sthitashchetasi drushyate II 38 II

Now after understanding MIND, in its reality, there is only one supreme existence, which is super soul. This beams with full consciousness, and from which all creations, preservation and at the dissolution is taking place. That is the meaning of creation from MIND (Chapter 10 stanza 6 stated above).

Some scientific facts are explained in various Sanskrit texts:

Madbhagwat & Devi Puran of Madbhagwat etc.

⅄ Embryo growth outside body – In-vitro.
⅄ Embryo transfer from one female to anotheremale.

Fetus from Ovum, Sperm and Ovum & sperm both.
Chromosomes GUNVIDHI in Sanskrit.

⅄ Heart beat in 2nd month of pregnancy.

Prashnopnishad & Shankaracharya

⅄ graitational force

ATREYA UPNISHAD.

Madbhagwat & Pudgal Shastra

⅄ Atom is divisible & independently exist.

Aryabhatt –knew Value of Pi =3.14 centuries ago
Modern science understood all this by 18th Century.

At the end of this chapter it is explained that super soul with its smallest fraction the whole universe is pervaded that is how everything is in control of super soul.:

Athava bahuneiten kim dnyaten tavarjun I
vishtabhyahmidam krutsnamekanshen sthito jagat II 42 II
(Gita stanza 42).

In summery the universe manifest and unmanifest world is due to supreme soul. It is generated and maintained by the process of automation. The facts are explained scientifically with quotations of Sanskrit stanzas from various available Sanskrit literature.

Before we go to next chapter 11, of Gita about "VISHVA ROOP DARSHAN YOG", basic understanding about our body, mind and intellect etc. seems mandatory for better acceptance of science involved in the stanzas.

In Bhagwat Gita chapter 13, stanza 6 and 7, body is SANGH or composite accumulation or conglomeration (Sanghatshchetana dhriti) of 24 elements:

Ichchha dveshah sukham dukham sanghatschetana dhriti I
etat kshetram samisen savikarmudahritam II 7 II
(Gita 13, 7).

It is must to understand the Gita word "SANGHATSHCHETANA DHRITI". The hidden science in this is about physics, the theory of relativity and quantum physics. Life lives upon death is the universal law. It is factually understood and accepted by scientists, philosophers, all religions, faiths, cults etc. Our body lives upon death. The destruction of food items releasing vital ingredients, particles, chemicals etc. One thing dies the other thrives. In microform electron moves around proton. The electron energy is due to its movement. This energy is "chetana". Combinations and reactions of electron energy, resulting into element compound formations as discussed earlier. This is responsible for existence in the form which is visible perceivable to sense organs. In fact, there is nothing. Or nothingness is responsible for existence. Everything is illusionary. Nasdiya sukta mentioned earlier, speaks the same. There is, was no truth or untruth, there was no medium of expression, nor space, no demise or immortality, no night nor day. There was only existence, pulsating within itself. One knows only this. All is illusion. Our body life is only idea. The death is end of metabolism. Or death is demise of conception. Whole creation is product of cultivation. Combinations and reactions are due to energy. Constant cultivations on energy is responsible for existence in the form of electron, proton, neutron, atom, Molecules. The sense of life is due to such energy or chetna. All is part of quantum of nothingness.

The physical body and vital energy in it is undergoing change every moment. In our body the replenishment of cells, tissues and structures is a continuous process. Skin is rejuvenated, within the span of 15 days, Red Blood cells every 120 days, muscles in 2 months, bones in 6 months and so on. In fact, physical body (or SHAREER in Sanskrit, it means "Sheeryate prati kshanam shareer" in BUDHA philosophy it is considered as "Sarvam kshanikam sarvam chanchalam"), and vital energy in the body both change every moment. As has already mentioned, the cultivation reorganizes new arrangement of quality particles in genes. *(Yogiraj Manohar Harkare "Enigma of death and Birth" VaidiknVishva publications, 1st edition, 1997, Printed at Mudrikka Graphica, Hyderabad.)*

CHAPTER ELEVEN

VISHVA ROOP DARSHAN YOG
ABSOLUTE TRUTH SUPEREME SOUL
SCIENCE OF TIME
(OR KAAL VIDNYAN)

Abstract: This chapter is devoted to Universal form of supreme soul – "VISHVA ROOP DARSHAN YOG". As per description of super soul beyond time (Kaalah traya teetah), which has no form (deha trayah teetah:), shape or no attributes (Guntrayateetah) then how a form possible to super soul? Atharva Ved describes soul as the soul is beyond time, form, and virtues. What lord KRISHNA described to Arjun is his form as super soul. With special vision provided to Arjun by Shri Krishn, universal form (VISHVA ROOP) was made visible to Arjun. Anything which does not have form or virtue or is beyond time, then how a form is made visible? Is not enigmatic?

In first stanza of this chapter Arjun queries to Lord Krishn that, though the illusion has finished, but how the supreme soul has entered into the cosmic form? Arjun wanted to see the cosmic universal form. The author of Gita trying to illumine the population through Arjun. In the form of Shri Krishna Author explains to Arjun:

Pashya me parth rupani shatashotha sahastrashah I
nanavidhani divyani nanavarnakrutina cha II 5 II

(Gita Chapter II).

Arjun now you see hundreds and thousands of forms of different type and colour of divine super soul. In stanza 6, different Aadityas, Vasus, Rudra, Ashvins, Marut etc. which no one has seen before. Details of these available in Sanskrit literature (Details are in Vaman Puran). Description is as under:

Pashyadityanvsunrudranashvinou marutastatha I
bahunyadrushtpurvani bharat II6 II

(Gita II, 6),

and further detail in stanza 22:

Rudraditya vasavo ye cha sadhya vishveshvinou
marutashchoshmapashcha I
gandharvayakshasurasidhasangha vikshante tvam
vismitashcheivasarve II 22 II

All these Aditya, Vasu etc. mentioned above are seen in VISHVA ROOP, and all including Gandharv, Yaksha, Demons, Illumined, (gandharvayakshasurasidhasangha) are wonderstruck. In Vishva Roop what was visible is narrated in this verse. Seen are many Aditya, Vasu, Rudra, Ashvin Marut etc. What is the scientific fact of all these things in VISHVA roop. Details about this is as follows:

Aditya are twelve in number. aditya are twelve Zodiacs, names are – Mitra, Ravi, Surya, Bhanu, Khag, pushna, Hiranyagarbha, Marich, Aditya, Savitra, Ark, Bhaskar.

Vasu (energies in and outside living beings and all manifestations) are 12 in number. These are powers or energies behind every manifestation. As thought energies with in us or energies outside us. Thought power can be used to control these mighty forces. Names of each Vasu energies are:

Dhara (-earth), Nala (-Agni-), Nila (neel-wind-), Aha (atmosphere-), Pratysha (pre-dawn light), Dyu (twilight), Soma (moon), Dhruv (pole star). The control power rests in our senses and sense organs (10 in number), along with Mind and intellect (Total 12 in numbers).

Ashvini (without respiration) are two. Two types of Pranayam (Bahya and Antra Kumbhaks). Split of word Ashvin is (A=no, Shvani=respiration), without respiration. Samadhi practicing Yogi can remain without respiration for long periods and remain disease free. This is supreme existence. Hence Ashvini are physicians of Gods.

Rudra (ferocious, devstations) are 11 in numbers. Rudra means ferocious, devastation etc. These are: 5 – abstractions (Anand, Vidnyan, Manas, Pran, Life), 5 – Shiv-(Ishaan, Tatpurush, Aghor, Bhairav, Sddyojat). One Atm or self.

Adityanamaham vishnurjyotisha raviramshuman I
marichrmarutamasmi nakshatranam Shashi II 21 II

(Gita 10. 21).

Meaning that, Among the Adityas (12 effulgent beings), I am Vishnu; among luminaries, I am the radiating Sun; among the Maruths (49 wind Gods), I am Marichi; among heavenly bodies (NAKSHTRA), I am the Moon.

Moon is not from Sun family. Vedik knowledge considers moon to be a NAKSHTRA (nakshatranam Shashi). **These NAKSHTRA or star formations i.e. like 27 NAKSHTRA, which belong to the Galaxy. Therefore, it can be considered that the moon is derived from galaxy. This needs targeted research.**

Marut are 49 in numbers. Ved have described 49 types of winds in nature. These winds are blowing in space or Aakash. Vayu is one of the five basic elements discussed earlier (Panch Mahabhuts).

Main Maruts are seven. Details given below:

Following statement is for interested reader in ready reference from 49 Maruts mentioned by Sri Sri Yogiraj Lahiri Mahasya Maharaj

49 Maruths:

vibratory element	Five pranas Place	Reigning God	Action	Location
½ Aakaasha	Prana	Visista	Crystallization	Heart
½ Vayu	Apana	Viswakarma	Elimination	Anus
½ Agni	Vyana	Viswayoni	Circulation	All over the body
½ Water	Udana	Aja	Metabolizing	Gullet
½ Earth	Samana	Jaya	Assimilation	Navel
Sub Airs	Sub Airs	Sub Airs	Sub Airs	Sub Airs
	Naaga		For belch	Gullet
	Krukara		to sneeze	Nose
	koorma		For movement of eyelids	Eyes
	Devadatta		Yawning	Mouth
	Dhananjaya		Keeps the body warm for ten minutes even after death	All over the body

49 Maruths are divided into main 7 chief category maruths. There are 7 sub class in each Marut. They are:

1. aavaha,
2. Pravaha,
3. vivaha,
4. paraavaha,
5. udvaha,
6. samvaha &
7. Parivaha.

Further description of these 7 is given below.

These maruths are existing within and without. That is the reason of relationship of existence in Cosmos, Mind, and Body.

The ever existing Brahma is expressed in 49 Maruths. Lack of this wisdom is leading us to ignorance.

The 49 Maruths are:

1. pravaha Swaasini taana mahaabal
2. Parivaha vihaga uddeeyaana rithavaaha
3. Parivaha sapthaswara shabdasthithi
4. Parivaha praana nimeelana bahirgamana thrisakra
5. Paraavaha maathariswaa anu satyajith
6. Paraavaha Jagath praana brahmaritha
7. Paraavaha pavamaana kriyaar paraavastha rithajith
8. Paraavaha navapraana praanaroopo chitwahith dhaathaa
9. Paraavaha hami moksha asthimithra
10. Paraavaha saaranj nitya pathivaasa
11. Paraavaha sthambhana sarvavyaapimitha
12. pravaha swasanaswaasa praswaasaadiindra
13. pravaha sadaagathi gamaadau gathi
14. pravaha pravadrisya sparsasakthi adrisyagathi
15. pravaha gandhavaaha anushna aseetha eedriksha
16. pravaha vaahachalavrithina
17. pravaha vegikaanthabhogakaama
18. udwaha vyaana jrimbhana aakunchana prasaarana dwisakra
19. aavaha gandhavaha gandher anuke aane thrisakra
20. aavaha aasuga saighram adriksha
21. aavaha maarutha bhittarer vaayu apaath
22. aavaha pavana pavana aparaajitha
23. aavaha fanipriya oordhwagathi dhriva
24. aavaha niswaasaka twagindriya vyaapi yuthirga
25. aavaha udaana udgeerana sakrith
26. parivaha anil anushna aseetha akshaya
27. parivaha samirana paschimer vaayu susena
28. parivaha anushna seethasparsa pasadeeksha
29. parivaha sukhaasa sukhadaa devadeva
30. vivaha vaathivyak sambhava
31. vivahapranathi dhaaranaa anamithra
32. vivaha prakampana kampana bheema
33. vivaha samaana poshana ekajyothi
34. udvaha marutha uttaradiger vaayusenaa jith
35. udvaha nabhasthaana apamkaja abhiyuktha
36. udvaha dhunidhwaja aadimitha
37. udvaha kampanaa sechana dharthaa
38. udvaha vaasa dehavyaapi vidhaarana
39. udvaha mriga vaahana vidyuth varana

40. samvaha chamchala uthkshepana dwijyothi
41. samvaha prushathaamapathi balam mahaabala
42. samvaha apaana kshudhaakara adhogamana ekasakra
43. vivaha sparsana sparsa viraat
44. vivaha vaatha thiryak gamana puraanahya
45. vivaha prabhanjana manapruthak sumitha
46. samvaha ajagath praana janma marana adrisya
47. samvaha aavak phlaa purimithra
48. samvaha samira praatah kaaler vaayusanj mitha
49. samvaha prakampana gandher anuke aane mithaasana

Division:

A) 7
1.   pravaha Swaasini taana mahaabal
12. pravaha swasanaswaasa praswaasaadiindra
13. pravaha sadaagathi gamaadau gathi
14. pravaha pravadrisya sparsasakthi adrisyagathi
15. pravaha gandhavaaha anushna aseetha eedriksha
16. pravaha vaahachalavrithina
17. Pravaha vegikaantha bhogakaama

B) 7
2.   Parivaha vihaga uddeeyaana rithavaaha
3.   Parivaha sapthaswara shabdasthithi
4.   Parivaha praana nimeelana bahirgamana thrisakra
26. parivaha anil anushna aseetha akshaya
27. parivaha samirana paschimer vaayu susena
28. parivaha anushna seethasparsa pasadeeksha
29. Parivaha sukhaasa sukhadaa devadeva

C) 7
5.   Paraavaha maathariswaa anu satyajith
6.   Paraavaha Jagath praana brahmaritha
7.   Paraavaha pavamaana kriyaar paraavastha rithajith
8.   Paraavaha navapraana praanaroopo chitwahith dhaathaa
9.   Paraavaha hami moksha asthimithra
10. Paraavaha saaranj nitya pathivaasa
11. Paraavaha sthambhana sarvavyaapimitha

D) 7
18. udwaha vyaana jrimbhana aakunchana prasaarana dwisakra

34. udvaha marutha uttaradiger vaayusenaa jith
35. udvaha nabhasthaana apamkaja abhiyuktha
36. udvaha dhunidhwaja aadimitha
37. udvaha kampanaa sechana dharthaa
38. udvaha vaasa dehavyaapi vidhaarana
39. udvaha mriga vaahana vidyuth varana

E) 7
19. aavaha gandhavaha gandher anuke aane thrisakra
20. aavaha aasuga saighram adriksha
21. aavaha maarutha bhittarer vaayu apaath
22. samvaha chamchala uthkshepana dwijyothi
40. samvaha prushathaamapathi balam mahaabala
41. aavaha pavana pavana aparaajitha
23. aavaha fanipriya oordhwagathi dhriva
24. aavaha niswaasaka twagindriya vyaapi yuthirga
25. aavaha udaana udgeerana sakrith

F) 7
30. vivaha vaathivyak sambhava
31. vivahapranathi dhaaranaa anamithra
32. vivaha prakampana kampana bheema
33. vivaha samaana poshana ekajyothi
43. vivaha sparsana sparsa viraat
44. vivaha vaatha thiryak gamana puraanahya
45. vivaha prabhanjana manapruthak sumitha

G) 7
42. samvaha apaana kshudhaakara adhogamana ekasakra
46. samvaha ajagath praana janma marana adrisya
47. samvaha aavak phlaa purimithra
48. samvaha samira praatah kaaler vaayusanj mitha
49. samvaha prakampana gandher anuke aane mithaasana

Each and everything of manifest and unmanifest world originating from and visible in Krishna's "VISHWA ROOP". Each and all appearing from him and disappearing back into him. What is the scientific fact hidden in this Vishva Roop? For reluctant to fight Arjun, Lord Krishna exposes to amazing form or Vishva roop, to make him realize that everything in material world is getting unmanifest back in super soul. To understand science in this scrutinization of Sanskrit literature the description is worth to ponder on KATHOPNISHAD, 1. 2. 24 and 25.

Asya brahma cha kshatram cha ubhe bhavat odanah I
mrutyuryasyopasechanam ka ittha ved yatra sah II 1. 2. 25 II

In the stanza previous, to this i.e. 1. 2. 24.

Navirato dushcharitannashanto nasmahitah I
nashantmanaso vapi pradnyaneneinamapnuyat II 24 II

Un-pious person whose senses have not been quietened, "mind set" is uncontrolled, even with unequivocal powerful knowledge, cannot achieve or get up to the soul and super soul. One knows that in a lake with water which is restless, reflected image of moon cannot be appreciated properly, similarly how one can understand self (or aatma) with restless mind set?. There is difference between knowledge about soul and get into the soul. The subject of stanza 25 above is different. Material and physical power both is food and death a drink for super soul. This means everything originates and vanishes back into super soul as in chapter 5 above. Scientific explanation is:

All manifest and un-manifest ultimately, return into "Him". All of material world comes out of 'Him' and take refuse into Him. Maximum part of this chapter has described about the Frightful, dangerous form of the super soul like monster. Each and everything in this manifest material world/universe, originates/manifests from and disintegrate/unmanifest back into the super soul. This needs time scale. Manifestation and expansion phase of universe after which regression and un-manifestation phase or ultimate annihilation of all the universe is within the frame of time span i.e. KAAL MAHIMA. As expressed earlier it is periodical and exhibits in a cyclical manner. It is a cosmic drama enacted by super soul. (Paramaatma).

To understand Time span and its importance, Gita chapter 14, stanza 5 is important. Time and region (desh, kaal tatva) element must be understood. Our vision is limited on this subject of time and region. It is outcome or manifestation of NATURE (prakriti).

Satvam rajastam iti gunah prakritisambhavah I
nibdhnanti mahabaho dehe dehinamvyayamII 5 II
(Gita chapter 14).

Three virtues are responsible for manifestation of Nature. When soul in connection with nature, it gets entangled with three virtues. Change or breakdown in these three virtues generates Nature.

Change in nature generates virtues. First manifests Goodness – (SATVA GUN), and then Passion (– rajo gun), Ignorance (tamo gun). Tamogun is responsible for the manifestation of five gross elemets of which Space or AAKASH is one (we are concerned here is time which manifests in aakash). The space is the main support of region and time. Time and region manifest from space and in SPACE. The ERA or YUG, Year, month and day all gross time manifest from NATURE. This ultimately unmanifest in super soul.

From the perspective of modern human understanding the world is in a crisis of moral and spiritual behavior. Populations all over the world seems frustrated and helpless. The science has overcome the barrier of time and space but has failed to promote better understanding between factions of humanity all over the world between nation and nation. Nations in modern life are facing distrust, hatred, violence. Humanity, globally trying to conquer nature but failing to conquer its own mind and senses (chitta, indriya). By transcendental meditation this is possible. But humanity has no time or devotion to this. Sages of BHARATVARSH through science of spirituality this aspect of humanity, had mastered perfectly. In modern psyche this campaign has already been initiated.

Material world, the cosmos and entire universe, is vast and endless, but the mind which tries to encompass using senses, seems bigger than the cosmos, as it grasps and comprehend cosmos. Control of such mind can result miracles. This fact was understood by Vedik Sages. Practice of YOGA, ASANS (POSTURES) can help in achieving mind control. But is not beyond time. Time controls everything in universe. But time is regulated by super soul and is under control of SHIV or super soul, as per VEDIK literature.

On realization by Arjun, that super soul form of KRISHN is the father of all this cosmic manifestation of moving and non-moving, spiritual master, and is worth worshiping, incomparable, immeasurable etc. Lord KRISHN showed absolute truth to him. Later on at the end of VISHVA ROOP form, the normal human form is displayed. This was to satisfy ARJUN and his query and restore his original nature as warrior. What science is in this description of "absolute truth".

Details are available in VED, PURAN, UPANISHAD, ETC. Following narration is to state what science is hidden in the literature of Sanskrit Speaking Civilization. Interestingly, sages of ancient India have struggled a lot to understand absolute truth about creation of universe. **After immense success they still were not sure of ultimate understanding about absolute truth, and have commented that, "thus so far we know beyond this you investigate and explore". The investigation is labelled as "Na Iti" or "Neti". This is not the end of investigation. There is still more to be investigated. What a scientific and truthful acceptance of facts. Modern science result mostly consider that everything has been achieved.** But the fact is otherwise e.g. Earth was considered as flat and later, proved to be round. Sun rotates around earth but now sun is stationary, and earth moves around sun. On the contrary Sanskrit speaking investigator summarized as "Na Iti or Neti".

Shri Krishn exposes to ARJUN, the functioning of absolute that everything of material world of universe which generates from super soul and merges back to super soul at the end, in a cyclical manner. – Super soul, self-manifest, is cause of manifest and un-manifest material world of universe. SHIV is considered as self-manifest, Vishnu is resting on floating serpent bed over "sea with a whitish hue" named KSHIRSAGAR and BRAHMA manifested from central part, umbilicus of VISHNU. This description is for common mind to understand the truth. Because for majority minds it is difficult to understand the real absolute truth hence this simple presentation.

But scientifically this can be explained as:

SHIV is self-manifest and is cause of all manifest and un-manifest universe. (**AITTARIYOPNISHAD** quoted in previous chapter – BRAHMOTPATTI). SHIV is in a form of AGNI-LING of FIRE, multiple blasts are taking place in it. It is like SUN – a ball of fire exhibiting multiple solar flares, Due to multiple blasts in AGNI-LING, gaseous balls are emitted repeatedly which on cooling contribute to the formation of galaxies and stars of universe. Effect of gases around forms layers of gasses around AGNI-LING – clouds of **primordial gases** which are probably of whitish hue like milk, hence termed as KSHIR SAGAR. – "milky sea" (KSHIR = white or MILK in Sanskrit). On this MILKY SEA, VISHNU is in lying position – i.e. sleeping posture. Scientifically in a DORMENT PHASE i.e. no active participation in BRAHM creation. But this primordial gas cover as VISHNU, protects from the harmful effects of radiation, harmful rays, extreme heat, and tej, light or prakaash of AGNI-LING, on material world of BRAHM which originates from hot gasses. In other words when harmful effects of rays and heat of SHIV is barred by gaseous layer, the BRAHM takes birth or manifests in due course of time. The brahm is manifested from the center of VISHNU the primordial gas cover (depicted as from umbilicus of VISHNU). The universe manifests and un-manifests **cyclically in a time scale** as mentioned earlier. The destruction is under control of SHIV at the time of annihilation. Thus, SHIV is destructor, apart from creator as well. Thus, VISHNU is protector and BRAHM is generator of manifest world. Details given below.

Science of time is perfectly described in Shivpuran as "kaalmahima" Glory of time.

In SHIV MAHAPURAN in Chapter of VAYVIYA SAMHITA while glorifying TIME (KAL-MAHIMA), the universe lives in cycle of time. Day of BRAHMA is KALP, each KALP has 14 MANVANTAR.

Each MANVANTARA has 71 YUG CHATUASK. This has been described earlier. Names of Manvantar which have passed out so far or ended are:

SWAYAMBHUV, SWAROCHISH, UTTAM, TAMAS, RAIWAT, CHAKSUSH

Current manvantar which running is VAIVASWAT MANVANTAR

Other manvantar which will follow current VAIVASWAT MANVANTAR are:

SAVARNI, ROUCHYA, BRAHMSAVARNI, DHARMSAVARNI, RUDRASAVARNI, DEVSAVARNI and INDRASAVARNI.

Thus 14 MANVANTARA COVERING 1000 years of YUG have been described. KALP

In SHIVMAHAPURAN it has been described that AGNI SHIV LING – was explored by BRAHM and VISHNU but could not find its end point anywhere of that limitless ball of fire – as SHIV LING. It is huge and limitless. Same as mentioned by ARJUN in BhagwatGita Ch. 11 stanza 17, after seeing **Divine Brillience form** or DIVYA ROOP of Shri KRISHN. SHIV LING of fire is very bright as if millions of suns shining. In sun, solar flares are taking place every now and then, emitting RADIATION, heat and light every time considered as nuclear blasts. Similarly in SHIV LING of fire multiple blasts taking place:

Kiritinam gadinam chakrinam cha tejorashim sarvato diptimantam I
pashyami tvam durnirikshyam samantaddeeptanalarkadyutima
prameyam II 17 II

(Gita 11, 17).

Because of immense bright light radiating from divine form Arjun seems unable to see the displayed form of Divya Roop. The glitter of immeasurable fire effulgence and radiations of suns. Scientific analysis is at the end of this chapter.

Lot of hot gases come out of it. Gas clouds are spread all around LING – WHITISH hue of PRIMORDIAL gases. Apparently looking like KSHIR SAGAR – sea of milk. The Vishnu keep rotating around SHIV LING OF FIRE in cycle repeatedly, from this primordial gases balls of gases thrown out forming groups of stars, galaxies on cooling. All moving in cycles around the central VISHNU. Thus, time is needed to complete the cycle. This period is considered as Day and year of Brahma and Vishnu and decides age/life of Vishnu and Brahma. So are the calculations described in Ved and Puran. After imagining all this one can straighten the events in universe. Days and years of Brahma and Vishnu have been calculated by sages of Sanskrit Speaking civilization. What NASA is seeing and calculating in USA is not within the reach of a common mind today. Modern day scientists have calculated that our present day universe is 155 billion years of age. Calculations by Vedik sages are also the same. BRAHM lives for 100 years. One year of BRAHM has 360 days. One day is KALP and same is night. Brahm or universe manifesting during day of Brahm and it unmanifests during nignt of Brahm. This is cyclically repeated during 100 years of life of Brahm.

Some details from TAITTAROPNISHAD are described in Chapter 10 (page 108-9) on formation of material world PRAKRITI. The embryological development compared to explain the manifestation and formation of material world in the sections of Brahmotpatti and prajotpatti. Indirectly this suggests the development of embryological science was well developed and was a common knowledge in general population. Shiv himself describes: "Who am I, where am I, How am I" (Title of a Marathi Poem (**MEE, KON, KUTHE NI KASA**)

MEE, KON, KUTHE NI KASA

Kshano kshani sada jithe divya pinda prajvalit
Bhavya divya niskhalit nistabdha mi vase tithe II

bamam bamam ninadita bhavya srshti prasadita
nirgun nirakarita alakh sarv jithe tithe II

anorani mi ase tithe mahatatva mi vase tithe
utpanna jag prasthapita sarvatra vas kare tithe II

prarabdha pind prasthapita svayam chalit jag sodita
nasoni ase bhas tithe abhasit mi vase tithe II

mahatatva prasthapita mahirupen pratishthita
asoni mi nase tithe tarihi mi vase tithe II

Bhase na surya Chandra jithe na prajvalit agni tithe
Omkaar shabda ninadita nistabdh mi vase tithe II

(SATYABRAHMA JAGANMITHYA by Dr. B.G. Matapurkar, Publisher Nandini Sanosh Tamboli, 42. Page 61. 2009, ISBN: 978-81-907405-1-7).

Author's poem from **"Satyabrahm Jaganmithya"**.

Super soul's abode is where every moment huge balls of fire are discharged in quietude I reside there. With BOOM sound I create Nature where I am without form, shape, virtue or with visible presence. With micro presence in smallest and biggest elements of material world and establishing created and automated universe, my presence is only experienced as my presence is only can be felt. On establishing biggest forms I am only in very microform in that. I am there but not there at the same time, but still I am present. Where sun, moon, fire loose their identity, sounding OM sound silently there I register my presence.

Another question of this chapter is how Arjun got special vision to see all this cosmic drama shown by Lord Krishna? Special vision was bestowed by Lord Krishna to Arjun. Similar vision was also given to SANJAY who narrates this cosmic drama to caretaker blind King DHRITRASHTRA of Hastinapur, sitting in his room. Partly this has been answered in Chapter 1 of this book. But scientific explanation could not be found in the literature scrutinized.

Another question is about the absolute truth. Why a human must know the absolute truth? Why Shri Krishn is impressing Arjun to know about absolute truth? Lord Krishn has explained that the human brain separates human beings from other living entities manifest in the material world. All living beings are influenced by the luring effect of MAYA or camouflaging influence of manifest material world. Sense enjoyment and sense gratification becomes the primary aim of all living entities. Sanskrit literature thought in a different way of liberation after enjoying life. Life's aim divided into Dharm, arth, kaam and moksh:

Dharm – of body is to maintain health, nutrition, proper function of body parts for enjoyment or suffering in life. Everybody thrives for it. It is basic 'dharm of body'.

Arth – is essential for food and necessities, enjoyment needs of body, and the society.

Kaam – is one which dominates the life may it be humans or other living entities. It included work and actions resultant effects of work and reproduction for maintenance of species in future.

Moksh – or liberation. Why and from whom the liberation is needed? Apart from dharm, arth, kaam as above the human mind-set ponders upon why "I" am here in this world? Where from "I" have come? Where "I" will go from here after death? What is my fate and fate of this universe? This pondering possibility is absent from brain and mind of other living beings of this material world. These questions separate the humans from other lives. These intriguing ponderings make humans special living entities. The SOUL or energy fraction of supreme soul is everywhere and in everybody.

Kalosmi lokakshaya krutpravridho lokansamahartumih prvrutah I
rutepi tvam na bhavishyanti sarve yevasthitah pratyanikeshu
yodhah II 32 II

(Gita II, 32)

In this Lord Krishna impresses upon Arjun that the absolute truth ultimately destroys everything, and same is reiterated in Kathopnishad 1, 2, 25 stated above. The display of Viashva Roop is to impress that the Absolute truth form is all devouring giant. That is even if Arjun refuses to fight, the absolute truth is going to devour everything in time by some way or the other. Fighting forces on both sides are destined to be slain. This is for the satisfaction of hunger and quenching thirst of death (mrutyuryasyopasechanam\).

Yasya brahma cha kshatram cha obhe bhavat odenah I
mrutyusyopasechanam ka itha ved yatra sah II 1. 2. 25 II

(Kathopnishad).

Even if Arjun refuses to fight and kill the enemy but as per the law of Nature all are destined to die/finish, by other methods of super soul. Death cannot be delayed or denied as per the law of supreme nature. Overall, this makes humans to know the truth of life and the material world. Lord Krishn tries to unfold these human ponderings by the knowledge of "ABSOLUTE TRUTH" (see Kathopnishad 1. 2. 25 stated above).

Scientific aspect of 6 and 12 stanzas of Gita chapter11:

Divisuryasahastrasya bhavedyugapadutthita I
yadi bhah sadrushi sa syad bhasatasya mahatmnah II 12 II

(Gita II, 12).

Modern science explains that if atom is split or break the atom a tremendous heat, light energy is emitted. Einstein proved it theoretically, but he could not split the atom. This was achieved by Oppenheimer (father of Atm bomb). The description after trinity test of the bomb is same as in Gita. Perfect Yogis have experienced such description as written in Gita. This points to the fact that the Sage Vyas who has written Gita had scientific knowledge and intellect. Divine universal form (divya roop) at one place but divided into many existences. Such a perception is above time and space (Gita 11, 13).

CHAPTER TWELVE

BHAKTI YOG – SPIRITUAL DEVOTION DEVOTION TO MANIFEST AND UNMANIFEST SUPER SOUL

This chapters focuses on devotion to formless or idol with form worship. The chapter discusses mainly on devotion or BHAKTI YOG. Bhakti means devotion. Arjun like any other human on earth, seems confused on the devotion to form less unmanifest (NIrgun Nirakar) super soul or manifest form (Sagun sakar form) idol worship – devotion to demigods. Which of the two is better.

Answering query from Arjun that which is superior Devotion (Bhakti Yog) i.e. worship of manifest God like Krishna or Devotion and worship of unmanifest god (Dnyan yog) like super soul, Lord Krishn explains scientifically that:

> *Mayyaveshya mano ye mam nityayokta upasate I*
> *shradhya parayopetaste me yuktatama matah II 2 II*
> **(Gita 12. 2).**

Focusing mind always in personal form with transcendental faith is considered as most perfect. Every action is in samadhi for personal form without any desire to result outcome. Let us understand this from science perspective.

Scientific pursuit is based on mind. Mind has been explained in chapter 10, with reference to Gita stanza 6, of chapter 10. Science needs evidence and proof. It can be grasped or felt mentally. Mind cannot grasp/see devotion i.e. unity of individual soul with super soul. Theoretically formless GOD is easy to speak but it is practically most difficult to perceive. Human mind structurally cannot conceive without form. Hence prayer or meditation becomes difficult. Therefore, considering the scientific aspect of mind, Sanskrit speaking civilization propagated "Idol" worship. This becomes Bhakti Yog, devotion for common population. Formless devotion or DHYAN yog is difficult to practice. Evolved humans get specialized into it. One can say Bhakti yog is first step to progress to dhyan yog. Stanza 6, advocates KARMSANYAS or renunciation of actions (that is giving up results of deeds and actions i.e. – KARMPHAL). The "Karmsanyas" is equally important in Sankhya yog, Dnyan yog, Karm yog, Bhakti yog etc. KARMASANYAS is action performed but not for self.

Space – Aakash, and the all – powerful nature, in fact is non existing. It is the mind which has the capacity to perceive it. Mind is supreme and infinite, and existing everywhere in all. In conclusion, the mind substance is the cause of nature and universe where our material body exists (Gita stanza 6, of chapter 10). In Yogvasishtha Ramayan chapter 71, stanza 64 it is very clearly illustrated:

Yasya hyatyantiko nashah syadsavuditah smrutah I
chittanasho hi nashah syatsa moksha iti kathyate II 64 II
(Gita 71, 64).

At the end of everything the end is in respect to mind. If mind gets destroyed it liberates the person from bondage and in turn person gets peace, wisdom, joy and gets exposed to absolute truth.

To understand the source of creation, let us further examine the different states of the mind. The internal mind gets exposed in three different states. Wakefulness-JAGRUT, sleep-SUSHUPTA, SWAPNA-DREAM. The wakeful state is difficult to accept as separate entity. At the end of this state when body relaxes the mind withdraws from all external connections. It gets absorbed within itself. Mind slips into state of sleep. Mind tends to forget everything to the extent even the existence of body also vanishes along with the world around. This is deep sleep or "Sushupti" state of mind. In sleep state sometimes one experiences another state called as dream state. Dream state resembles wakeful state. But the wakeful state gets falsified when person comes back to wakeful state. In other word dream state is the creation of mind. Everything appears real till falsified. Everything is created by mind from mind itself. A new world is created for mind itself. The mystery of dream state needs to be unearthed to understand the mystery of creation of nature and the universe. **After understanding all three states of mind, one realizes that there is only one supreme existence which points to consciousness, and from this consciousness the power of creation and manifestation comes to existence.** Mind is basically CHITTA in Sanskrit. Intellect, ego, will etc. are all representations of mind. The CHITTA or consciousness is recognized by different names like "I", Aatm or self. 'I' in gross form is body. The gross form of consciousness like body are nature and world. The names are different depending on functions of mind, but mind is one. Mind is in living body. It can be said that body functions only when mind operates in the body. Without mind, symptoms of life are absent. Or body is animated by mind. Mind supports life. Wakeful and dream state the parson is one and same. In conclusion the evolution of dream state is exactly is the same as wakeful state. The consciousness, the subtle becomes gross and creates world and universe is just an inference.

There are two different methods for BHAKTI YOG as per stanza 9. If one is unable to focus or meditate on super soul, a regulated continuous practice and study on super soul is essential.

Super soul is timeless or beyond time (KAALTRAYATEETAH), formless   (DEHTRAYATEETAH)   and   virtue   less

(GUNTRAYRTEETAH). This is as per ATHARVAVED. Existence of super soul can be experienced circumstantially. This can be experienced like wind/air, existence of which has no direct proof but indirectly it exists? Indirect evidence is also a proof. You may not see a distant fire as place where fire is, it is far away from your site but smoke erupts and goes high in the sky, which can be seen wallowing up in air from fire site, and can be seen from distance. After seeing smoke one can imagine that fire has erupted somewhere. Devotion is unity of spirit – soul with super soul. So is spirituality. Devotion can be personal or impersonal BRAHM i.e. devotion to super soul, which is un-manifest, invisible, un-conceivable, having no attributes and no shape. Science is based on evidence that does not mean if perception of evidence is absent, science does not exist. This led Sages and Rishis to project different types of evidences. *Examination and analytical study of material element with the help of different evidences is NYAY – a fair judgement.* Accordingly four main evidences described in NYAY DARSHAN by Rishi Gautam (personal name = AKSHAPAD). The names and types of evidences are in the stanza 1 of chapter 1 of NYAY DARSHAN:

Pramanprameyasanshayaprayojandrishtantsiddhantavayavatar kanirnayavadajalpavitandahetvabhaschhalajatinigrahsthananam tatvadnyanannihi shreyasadhigamah I

(NYAY I, 1.)

Resource of true knowledge is "Evidence". Main PRAMAAN are four in number: 1. Direct evidence (PRATYAKSH). 2. Indirect/estimated evidence (ANUMAAN). 3. Similarity evidence (UPMAAN). 4. Incomprehensible evidence (AAGAM). (This is a brief statement. For details original work be consulted)

Personal devotion is where transcendentally attached to a "particular-form" of GOD e.g. KRISHN. This is BHAKTI YOG. It is easy to achieve by meditation transcendentally. On the other hand, impersonal or un-manifest form of super soul, is difficult and un-conceivable.

Transcendental attachment to impersonal form of BRAHMA is known as "Yog of knowledge" or DNYAN YOG. The impersonal devotion is troublesome, difficult and needs a lot of patience to achieve goal. One can practice any of the technique, ultimately supreme soul form of Brahma is achieved by both i.e. Bhakti yog and Dnyan yog techniques. BHAKTI yog is easy and direct devotion service to super soul –BRAHM. Worship of GOD in temple, is meditating on supreme soul as idol represented with material qualities. It is SAGUN worship. Ultimately the aim, is worship of BRAHM. Spirituality for self-realization is not to be practiced through difficult and uncertain path because the result may be uncertain. Hence GITA stanzas advocate to follow certain path for better result which is BHAKTI YOG. The achievements are possible easily by attempting Bhakti Yoga for this simple method is advocated. Invest mind in super soul, renounce results of deeds/work, always remain satisfied in dualities of life, like grief and pleasure, loss or gain, success or failure, fame or defame etc. live without fear and anxiety, don't be disturbed by auspicious or inauspicious happenings, equally respect friend or enemy, equal feeling for praise or criticism and with fixed determination trust in super soul, such a determined person is very dear to super soul (Lord Krishn):

Yasmannoudivajate loko lokannodvijate cha yah I
harshamarsha bhayodvegeirmukto yah sa cha me priyah

(GITA, 12, 15).

Ye na hrishyati na dveshti na shochati na kangkshati I
shubhashubhaparityagi bhaktiman yah sa me priyah

(GITA 12, 17).

GITA advocates to practice spiritualism in such a manner with faith, attachment and devotion to service to super soul. With such devotional practice Mahatma Gandhi could uproot the deep rooted mighty British Empire from INDIA, with peaceful but firm resistance. But the role of other freedom fighters cannot be belittled.

Life energy transforms into physical material human body. Similarly, life energy gets different bodies. That means different forms of body the life energy is one and the same. This is DHARM of life energy. When one sees such different forms then it is assumed that the bodies are life itself. It feels like as human form is superior to all other living bodies. With such egoistic mind set, humans develop ego. Man treats the other living bodies as low category and ill treats them.

What is Super soul is explained in Katopnishad 2. 1. 9, indirectly as:

Yataschodeti suryostam yatra cha gachchhati I
tam devah sarve arpitastadu natyeti kashchana I
etadave tat II 9 II

From where the SUN is generated and where SUN merges after annihilation and all gods (elements of material world – Panchmahabhut) ultimately merge, beyond this nothing can reach, is super soul. He is Omniscient and Omnipresent. Bodily forms can be different, but He is same in everything in the universe. It is unwise to consider superior or inferior. Such Mind Set will certainly solve many problems of material world. This is explained with illustration with physics principle in Kathopnishad.

Yathodakam durge vrushtam parvateshu vidhavati I
evam dharman pruthak pashyam stanevanuvidhavati
KATHOPNISHAD Chapter 2. Section 1. Stanza 14.

Rainwater fell on top of a peak of a mountain, flows down in different streams. Similarly living entities see other living being as different bodies. Even then the soul (Aatma-आत्मा) in different bodies is same. Man considers other bodies as inferior and treats them as inferior thinking himself as superior. Resulting into evil actions towards these bodies. Hence the result of such actions experienced by that person. The fraction, of Super soul, has to undergo life in such different bodies (stanevanuvidhavati). This is as per the principle of Physics. Waves in sea look different – have different shapes (small or big) but water is same. When a wave comes to the shore it returns. A drop of water may be on top of a wave and other may be deep. On top is man and deep may be any other life form. But in other phase of wave the position may be different. Drop on top may get deep and deep position deeper one may come to top. Similarly, in life forms may take different positions. The waves in sea are on top part of sea. But not in deep sea. Consider the life form in material world of BRAHM, the waves do not remain in deep part of BRAHM. Different bodies in life form do not remain separate. They are at par with BRAHM. All this is beautifully described in Kathopnishad:

Yathodakam shudhe shudhamasiktam tadrugeva bhavati I

evam munervijanat aatma bhavati goutam

(KATHOPNISHAD 2. 1. 15)

Pure water when poured in pure water it becomes the same. Similarly, MUNI when understands this the soul of such a muni becomes 'Brahm' itself.

The super soul (or Paramatma) is like a "flame without smoke". Because a flame with smoke is not pure fire. HE is JYOTIRMAY TEJ. HE is pure light and knower of past, present and future as stated in Kathopnishad 2. 1. 13:-

Angushtha matrah purusho jyotirivadhumakah I

ishano bhutabhavyasya sa evadya sa u shvah I etadve tat II 13 II.

The same is scientifically explained in stanzas of GITA Ch. 12. Stanzas 15, 16, 17. Ignore all dualities of life (as stated earlier) to be happy and carefree in life.

The techniques advocated in devotional yog, can keep individual stress free and tension free in all walks of life. Many social and psychological problems can be obviated by adopting such techniques. This is highly "scientific advice". This needs to be adopted by Medical practitioners in the treatment of modern Psychological problems. In other word, it is a treatment in many psychological ailments and diseases due to stress and strain of modern life.

Ye tu dharmyamrutamidam yathoktam paryupasate I
shradhanamatparama bhaktastetiv me priyah II 20 II
(Gita 12, 20).

Follower of the path of oneness with everything and every living being, with devotion as per vedik teaching, is dear to Super soul.

This is the science of tendencies explained in 12th chapter. The eternal immortal DHARM of Sanskrit speaking civilization does not discriminate between any religion, faith, cult or Sampraday – Hindu, Muslim, Christian, Parsi, Buddhism, Jainism, Sikhism, Jewism etc. which are artificial faith compartments of societies. Sanatan dharm is for the elevation of incumbents of societies of humans of the world. This is Bhakti yog principle of the Vedik culture.

CHAPTER THIRTEEN

KSHETRA-KSHETRADNYA YOG NATURE, SOUL AND SUPER SOUL SPIRITUAL BIOLOGY

Abstract: Physical existence and owner of physical existence has been discussed in this chapter. Greedy mentality and tendency is harboured by us during our lifetime. Gita and Yog Vasishtha both trying to imbibe in us good tendencies so that to create good human society for the betterment of the whole world. The soul (jeevatma) functions through material bodies. Tendencies change the fine particles in DNA of our chromosomes. Clair science and the clair subject has been dealt in this chapter. The spiritual biology has been emphasised.

On enquiry by Arjun regarding PRAKRITI and PURUSH (nature and spirit), KSHETRA (field –the body), and Knower of KSHETRA the KSHETRADNYA (Soul, spirit, Aatma) the knowledge and the purpose of knowledge, Shri Krishna explains:

Idam shariram kounteya kshetramityabhidhiyate I
etadyo vetti te prahuh kshetradnya iti tadvidah II 2 II

(Gita 13, 2).

Body is field – material nature, (KSHETRA) the knower of this field is soul (ATMA) entrapped in material body. Body undergoes changes throughout the life span from child to youth or old age this is governed by super soul. It is automatic process. The soul remains unchanged. Nature or material body of nature is provided for enjoyment of soul called as – KSHETRADNYA.

The body is compared with field in Gita. Field simile, is an intelligent comparison full of wisdom, mentioned in Gita and Ved. One knows that if field is left to its own it may grow anything as per nature. Yield may be good or useless crop, even poisonous or harmful intoxicating crops, for human use, having good or harmful effect on humans. But properly cultivated and nurtured field, it yields useful crop. Similarly, if body is cultivated and nurtured it will be useful to society as well as all living beings on this earth. Khetradnya who knows the kshetra or field, – the soul – or mind has control over the field, can yield good or evil outcome. All depends upon the cultivation effect. Therefore, the Vedik civilization after realization of this fact devised SANSKAR method on human minds for the betterment of human societies.

Kshetradnyam chapi mam vidhi sarva kshetreshu bharat I
kshetrakshetradnya yordnyanam yattjdnyanam matam mama II3 II
(Gita 13, 3).

Knowing material body and the soul of different varieties of universe, is knowledge opines Lord Krishn

The body of material world is essential for field of activity of entrapped soul for enjoyment and hence the body is called as field – KSHETRA. The one who knows the field is KSHETRADNYA – Knower of the field. What is the hidden science in this statement of Gita? Detailed scientific knowledge of KSHETRA or body and KSHETRADNYA or soul of body can be explained from other sources of vedik literature as UPNISHADS.

In 6th chapter it is clarified that the thought process of Yogi, undergoes rearrangement of quality particles of DNA into more higher qualities and achieve divine and spiritual evolution. The quality particles get rearranged in each cell, as per evolutionary stage. This means that the material affinity gets transformed into vital form as KSHETRADNYA or soul and the soul into Kshetra or body or cells of the body. It is difficult to accept such an interconvertibility but on pondering and meditating on it this knowledge gets revealed. One must shun off ignorance on illusion of his mind. To reiterate this the 8th chapter narrates that Lord Krishn, himself is nature or prakriti or the body-Kshetra or field, and HE himself is knower of the field – kshetradnya-soul-jeevatma etc. The names are different and create confusion but meaning is same. Does not that mean the nature is just a stage of Super soul? That support the interchangeability.

This "atmatattva" i.e. soul element, a fraction of super soul, is called KSHETRADNYA. This ATMATATVA is complete in itself. It is a part of the super soul or BRAHMATATVA which is complete by all means. **Complete only can give rise to complete.** In other word, an incomplete element is unable to produce complete. The energy is present in moving and nonmoving fractions of living beings. Kshetra or body is composed of building blocks or multiple cells. In modern science it is known that the basic main element in living life is DNA molecule present in the cell. Deoxy ribonucleic acid (DNA) which is again a complete element. It can produce another complete DNA molecule. Living entity is also produced from complete ATMATATVA. Life is a living cell and cell can only produce cell (Virchow-http:/bio.libretexts.org). Living beings (PRAJOTPATTI) are produced like this only. Plant life is also produced like this. A primordial or 'stem cell' when forms another cell the original cell remains perfectly normal is a fact known to cell biologists. Regenerative, power of cell, depends upon the capacity of cell to divide and form similar new cell. In multicellular organisms, primordial cells, or STEM cells, have the capacity to divide. During divisions, one cell is preserved as original cell (as stem cell and remain in tissues and organs for repair and regeneration purpose of damaged tissues during life) and the other is left for specialization.

These specialized cells are responsible for tissues or organ formation of specialized structure. The different structure, of tissues and organs, are for different functions of such tissues and organs. On fertilization of ovum with sperm, the single cell ovum develop tremendous capacity to form complete animal in multi cell organisms. Stem cell of a single celled fertilized ovum of developing embryo form whole body having organs and tissues with different structures and different function. Thus, the whole human or animal body is formed. This is because of the completeness of the cell. In other word, it can be said that the soul element, (Aatmatattva) of the body or KSHETRA, the primordial stem cell itself is complete. Complete only generates, another complete cell.

(Matapurkar et all. American Society for Artificial Internal Organs Journal, ASAIO journal, 2003, DOI 10.1097.NAT.0000045043.61577.59.)

After certain numbers of divisions of cells, the division capacity of stem cells stops and the regeneration capacity of the body or KSHETRA is lost, resulting into old age of the body. The knower enlivens and energizes the body (or Kshetra) is soul energy or KSHETRADNYA or Aatmatattva or soul. Therefore BRAHM NATURE (Brahma Prakriti) is explained in ISHAVASYOPNISHAD (prayer Mantra):

Om purna madah purnamidam purnat purna mudachyate I
purnasya purna maday purn mevav shishyate I

Brahm is complete here and in the whole universe, it is complete, the complete dwells in the complete, even if complete is removed from complete the remaining is also complete. What an enigmatic statement is this? After understanding the Kshetra and Kshetradnya the RISHI and SAGES went further to analyze that Life is not for the body to live for sense enjoyment and gratification only and die. This body is provided by nature with fraction of energy from the whole universe – Super soul or BRAHMATATTVA.

Etat dnyeyam nityam eva aatmasanstham na atah param
veditavyam hi kinchit I
bhokta bhogyam preritaram cha matva sarva proktam trividham
brahmametatII
(SHWETASHWATTAROPNISHAD, 1. 12.)

The highlighted part of stanza means BRAHM has three form (**bhokta bhogyam preritaram**):

1.   nature or PRAKRITI, which is domain of work, – body. (**bhokta**)
2.   Physical nature or BHOUTIK PRAKRITI (**bhogyam**) and
3.   controller of nature or life energy – the soul (**preritaram**).

All three are part of super soul. In other word, the cell and the DNA of cell are controller of cell and through cell the body. The cell is workable through automation and supervision of KNOWER of body or KSHETRADNYA or soul.

Therefore, understanding of spirituality is also a great scientific job to be performed. Hence every time a great stress is enforced in gaining knowledge on spirituality, the VED and PURAN, which are full of such knowledges and needs to be followed. Accordingly, Shri Krishna explains to Arjun about KSHETRADNYA i.e. one who knows the "nature – given" body or material body. In this material body the soul is entrapped which enjoys the KSHETRA – the field or the body. Krishna explains that, "I am also a knower in all the bodies". To understand this, knower of body and the knower of all the living bodies of universe is understood as knowledge. (Chapter 14, Shlok 3).

Mamyonirmahad brahma tasmin garbham dadhamyaham I
smbhavah sarva bhutanam tato bhavati bharat I
(Gita 14, 3)

In VEDIK literature it is DNYAN. For this one has to know the PRAKRITI, the nature, the field (KSHETRA) which is body and Knower of the field or enjoyer of body – the field (KSHETRADNYA), and the knower of all the bodies – super soul (PARAMATMA). In other word, there are two souls in a body a local which knows only the body and limited to body only, and a universal soul which is a knower of all the bodies of universe which is controller of all. Similar expression is recorded in SHWETASHVATAR UPNISHAD. The BRAHM in the form of PRAKRITI or workplace (KARMKSHETRA), BRAHM in the form of life which controls physical/material body and the BRAHM which controls these two. It is explained in stanza above (Shwetashwatar upnishad 1. 12).

Details are given elsewhere in MADBHAGWAT and MARKANDEY PURAN, DEVIPURAN MAHABHAGWAT (SHAKTIPEETHANK).

Body is made from five elements earth, water, fire, air and space. These are PANCH MAHABHUT. Five elemental bodies of living beings are described as of 4 types.

Andajah swedajashcheivodibhajjashcheiva jarayujah I
Andajah pakshisarpadyah swedaja mashakadayah II 6 II
Vrukshagulmaprabhrutayashchyodbhijja hi vichetana II 7 II

(DEVI PURAN MAHABHAGVAT SHAKTIPITHANK:-Year 79, No. 1. Gita press Gorakhpur. Chapter 17, page 157.)

1. Born from EGGS (andaj) are birds and snakes,
2. Born from Larvae (swedaj) are Insects and Mosquitoes,
3. Born from seeds(udbeeja) are plants and vegetations,
4. Born as body covered with membrane (pandaj, jarayuj)

Apart from this it is further explained that "jarayuj" or born with membrane are from union of sperm and 0vum. These can be of three category: sperm (shukra/) excess – are males, ovum (rajah) excess – are females, and equal SHUKRA+RAJ produce third gender-Eunuchs (napunsak). This proves XX, XY theory of modern science. 9th stanza from Devi Puran is a testimony to this scientific statement:

shukradhikyen purusho bhavetpruthvidharadhipa I
raktadhikye bhavannari tayoh samye napusankam II 9 II

(DEVI PURAN MAHABHAGVAT SHAKTIPITHANK:-Year 79, No. 1. Gita press Gorakhpur. Chapter 17, page 157.)

Body is of 24 elements as per 6th and 7th stanzas of 13th chapter.

Mahabhutanya hankaro budhirvyaktamev cha I
Indriyani dsheikam cha pancha cheindriya gocharah

(GITA, 13,6).

Five Mahabhut i.e. Earth, water, Air, Fire and Aakash along with unmanifest three modes of mind, intellect, ego, five senses and five sense organs, five work organs, mind-set is internal sense organ. The total turns out to be 24. The summery of all this is given below.

Ichchha dvsh sukham dukham sanghatshchetana dhruti I
etat kshetram samisen savikarmudahrutam II 7 II

(GITA, 13,7).

Details as per above stanzas of GITA the different 24 elements have been described earlier.

Interestingly the common people of, vedik civilization were aware of all this knowledge. Peoples were knowledgeable with the scientific know how. The best testimony of this statement is the Lord Mc Caulay's statement in British Parliament in 1835. (Chapter 3, page 103) Five **great elements** earth, water, fire, air and ether in space (MAHABHUT), false ego, intelligence, un-manifest stage of three modes of nature, and five senses to **acquire knowledge** – eyes, ear, nose, tongue, skin. There are five **working senses** – voice, legs, hands, anus and genitals. There is sense within or **inner sense** the Mind. The **objects of senses** are five – smell, taste, form, touch and sound. All these are called as the **field of activities**. KSHETRA or field of activity or body is representation of these 24 factors. The 24 elements are encasement of soul which is entrapped in material body.

Twenty four elements conglomerated together in the form of material body is a potential vitality. Vitality (or chetana) is workable by the presence of energy or soul. From the physics point of view, theory of relativity of physics, the atom enlivens due to combined effect of its elements. Electron under relative coordination of time, space and other factions of atom, the life is created. Chemical actions and reactions are existing but occur when relative electrons coordinate in combination. Life is a product of combined effect of all elements. In summery all elements and body interact. To understand in a different way, one knows that the sperm and ovum are capable, of living independently. When unite form body but unable to exist separately without body of the mother. It cannot live long independently, unless the JEEVATMA or soul takes over by 7th month of pregnancy. Combined cooperation of mother's body and JEEVATMA the fetal body functions normally, in the uterus. This is SAMASEN-regular functioning of foetal body is cooperative endeavour of mother's body and JEEVATMA.

How to get out of it, is knowledge and the rest is ignorance. This has been described in GITA. Body (KSHETRA) is not permanent, but (KSHETRADNYA) the soul is. The soul is permanent and eternal. The inner sense is eternal. It is "conglomerated effect" of all senses of material body. State of mind is CHITTA (as no appropriate word for this it is used as it is. "Mind set" is CHITTA) which has control over mind and sense objects. Notifications reach brain centers from sense organs where these are analyzed and co-related for the knower – the soul. Soul is not PRAKRITI, or nature or material body. It is a fraction of superior energy – supreme soul. Do these natural sensations become supernatural on reaching soul? In fact, the senses are for smooth and un-hindered life till life span of field (body) or nature. The enjoyer decides to enjoy or renounce the joy. Mind is always busy in thought, planning, ambitions, past, present and future possibilities. For tranquility of mind it needs diversion to other modes. The meditation can change state of mind –the CHITTA and in turn control the state of mind. Meditation can be on any material entity, in the form of idol or material imaginary god or supernatural or super soul. This may probably the idea of "idol" worshiping depending upon the individual mind development, intelligence of the person. Deep engrossment of mind is BHAKTI or devotion as explained earlier. A way of diversion of routine business of mind. The essential of actual advancement in spiritual science is not for honor from others, such an acknowledgement which is false as the material body and its acceptance with oneself is a false ego. To understand the soul and body as a temporary entrapment of soul. It is not real and hence such an understanding is realization of real ego. I am BRAHM/spirit i.e. AHM BRAHMASMIN (BRIHD ARANYAKUPNISHAD"). Brahma is beyond the cause and effect of this material world. Scientific aspect of this apart from liberation, is 24 elements which make body or field, is to understand the different factors of which the body is formed. These 24 factors which entrap soul can also help in liberation of soul. (This body formation has already been explained scientifically earlier chapters in Prajotpatti section) It is like thorn can take out thorn or poison can cure poison. Instead of sense gratification taming of senses. Stanzas from 8-12 in 13th chapter of GITA clearly analyses pious living and self-realization. Behavioral pattern of an individual can be supervised and controlled for the good of the society and the humanity at large. This practice certainly lead to stress and tension free life which keeps body – KSHETRA disease free and healthy. Healthy body is capable of sense enjoyment and attempt at liberation to super soul. To tame the elements one must practice, Humility, non-violence, tolerance, simplicity, cleanliness, steadiness, and renunciation. All this can keep the body – the field suitable for cultivation, fit for everything in life. One cannot be fully happy or fully miserable. In all circumstances should remain equipoised. If one fails to understand the soul or super soul due to ignorance, he is slave or subordinate to material world and material nature. If one gets convinced about this truth, he realizes absolute truth. He is subordinate to supreme lord.

Sarvendriyagunabhasam sarvandriyavivarjitam l
asaktam sarva bhuchcheiva nirgunam gunabhoktu cha lll5ll
(Gita 13, 15).

Super soul is source and creator of sense but is without senses. He maintains nature. He is master of all nature and living and non-living material world. In the body there is another enjoyer – the soul, and one who understands the interaction of the two, can attain liberation.

Knowledgeable person can see the difference between body-KSHETRA and soul-KSHETRADNYA. One can understand liberation from bondage of material world thus can attain crmevcsupreme soul. Understanding that the body is matter, it can be analyzed with its 24 elements. Body works in collaboration of 24 elements and the soul under supervision of supreme soul.

The owner (soul – KSHETRADNYA) of Kshetra – field, using proper cultivations – SANSKAAR, can be motivated in life. It has been explained earlier. Hence proper SANSKAR needs to be administered for elevation to higher status of aim of life (Dharm, Arth, Kaam and Moksh).

Gita in its stanzas from 8th to 12 describes the techniques of proper cultivations for superior existence.

In chapter 13, the 16th stanza is interesting but difficult to understand. It can be scientifically anlyzed.

Bahirantshcha bhutanamcharamev ch l
sukshmatvattadvidnyeyam durastham chantike cha tat lll6 ll
(Gita 13, 16).

Outside and inside of all living beings whether moving or non-moving being in micro form or as subtle in nature, and unknowable as well with material senses, is far, far, away and near too

How is it possible? It is inside and outside of all living beings, far away and near, Un-knowable and subtle are all contradictory statements. How scientifically it is practical?

When one ignites a wood, smoke comes out and wallows up in sky, but its roots are at wood. Similarly, the subtle soul keeping roots at body can go up in space around. It is lighter than smoke hence can go longer than the smoke. The radio waves from radio station spread in surrounding space. It can be tapped anywhere on a radio-set. The sound waves are at the radio-set at the same time the waves are at the station and everywhere. So is the soul it is near as well as away. Therefore, the meaning of stanza is scientifically correct. As previously explained that the soul can travel anywhere in the space, leaving the body.

In Kathopnishad Chapter 1, stanza 21 similar meaning statement but in a different way has been described:

Aasino duram vrajati shayano yati sarvatah I

kstam madamadam devam mdnyo dnyatumrhati II2I II

(Gita 1, 21).

While in a sitting position still roams around in space, in sleep too HE can move everywhere. Full of pleasantries still without joy, such GOD – the Soul is known to me!

This statement is from GOD of Death (YAM-). Kathopnishad is from KRISHNA YAJURVED. It has two Chapters each having three VALLI. A person named VAJASHRAV had a son named NACHIKET – a saintly character. VAJASHRAV donated all his collection of old cows. Donation of useless cows is meaningless. NACHIKET did not like this donation. Hence, he asked his father that to whom are you donating me! On repeated asking father got irritated and in disgust said "to GOD of Death –". After thinking philosophically NACHIKET went to GOD of death. Katopnishad is in a dialog form between NACHIKET and YAM (God of death. For details refer to original work).

Stanza 21 explains YAM's statement to Nachiket. YAM knows the soul better because YAM very often deals with soul. Hence this statement – such GOD – the Soul is known to me! (devam mdnyo dnyatumrhati).

Interestingly Kathopnishad further says through statement by GOD of death (YAM) that, only listening, reading studying or only philosophizing on the subject one can not understand SOUL, Super soul or supreme soul. One has to wholeheartedly surrender himself to the soul. One must act accordingly with un bias mind. Thus, stanza says:

Aaymatma pravchanen lbhyo na medhaya na bahuna shruteyam I yameveisha vrunute ten labhyah tasyeish aatma vivrunute tanusvam II 23 II

(Kathopnishad 1, 2, 23).

There is difference between knowledge about soul and get into the soul and be soul. Whose food is both (brahma cha kshatram.), power of intellect (BUDDHI BAL) and power of physical body (SHARIRIK BAL), and drink is death (mruturyasyop sechanam), where is such entity? who has the capacity to know such an entity?

Yasya brahma cha ubhe bhavat oden I mrityuryasyopsechan ka itthya ved ytra sah II 25 II

(Kathopnishad 1, 2)

CHAPTER FOURTEEN

GUNTRAYAVIBHAG YOG
THREE VIRTUES (MODES) OF BRAHMA (NATURE)
SOURCE OF LIFE – A SCIENTIFIC PERSPECTIVE

Abstract: Three modes of nature are described in this chapter. Hence it is 'guntraya yog' or three virtues of brahma or nature responsible for creation of universe. Guna in detail are discussed. After manifestation of nature and formation of living body, explained in previous chapters, "Modes of nature" are discussed. In this chapter Lord KRISHN explains different modes of nature, and how modes act and how bondage and liberation is executed. Living entities are not born at the time of creation of universe nor disturbed at the time of annihilation. One can infer that the living beings are not born at the time of creation of nature. Living entities are created later part of evolution. Lord Krishna explains Arjun the knowledge and wisdom to attain supreme perfetion.

Idam dnyana mupashritya mam sadharmyabhagatah I
sargepi nopjayante pralaye na vythanti cha II 2 II

(Gita 14, 2).

Second line of the stanza expressed that Living beings are neither born with manifest nature nor die with annihilation of material world. As explained in earlier chapters living beings are developed from "Aap" or water after incubation by super soul and not at the time of creation of nature. This is as per modern science.

Total material substance is called BRAHMA. Supernatural soul impregnates (garbh dadhyamham\) this BRAHMA to generate all living beings as explained in the stanza:

Mam yonirmahad brahm tasmin garbh dadhamyaham I
sambhavah sarva bhutanam tato bhavati bharat II 3 II
(Gita 14, 3).

That is combination of Field-nature and knower of field the soul or body and spirit results in manifestation of living beings (Refer prajotpatti). The physical forms of material world (which is BRAHMA), are main source of life in whole universe. Physical nature and spiritual nature both are essential parts for birth of living being. Brahm is responsible for manifestation of all material world. The same has been mentioned in the MUNDAK Upnishad.

Yah sarvadnya sarvavidyasya dnyanmayam tapah I
tasmadetad brahm nam rupamannam cha jayate II 9 II
(MUNDAK Upnishad Ch. 1 section 1, stanza 9).

Omniscient, omnipresent and all knowledgeable BRAHMA with TAPAH the brilliance (TEJ), from whom this form and food is manifested.

The 5 elements have form i.e. have ROOP and are visible, are named BRAHM. Along with this "food (ANNA) BRAHM", is also manifested.

In RIGVED it is mentioned that the birth and existence of life in the universe is from SUN. The sun, is responsible for PRANIK energy and LIFE energy and because of such collective energy is recognized as MITRA or friend and VAISHWANAR –

Devanam karmapruthkvat pruyag abhidhan stutayo bhavanti I
ekeiva va mahanatma devata sa surya ityachakshate I
taduktam rishinam surya aatma jagatastasyushashcha
tadvibhutayo anya devatah

Vedic literature is full of knowledge about nature, soul or super soul. Natural elements are mentioned as god –DEVTA, special virtues and work of that god as mentioned above, decides the name for that god. MITRA DEVTA'S virtues like PRANIK energy and life force etc. as it has collective energy, SUN is named as VAISHWANAR.

Vishvan naran vibhatim iti veishvanarah I
vishvaschasou narashcheti vishvanara eva vishvanarah - sarv
pindatma ityrthah

All different types and forms of life are possible to take birth in this material nature. The super soul provides seeds and fathers life in universe. Types of life forms already grouped in previous chapters. In other word Physical/material and spiritual both are required for life to take birth and exist.

Satvam rajastama iti gunah prakriti sanbhavah I
nibandhati mahabaho dehe dehinamvyayam II 5 II

(Gita 14, 5).

Satv, rajah, tamah the three virtues are the creation of nature or prakriti. The soul is confined with these three virtues, in the physical material body. Pleasure, dire desire for work action and greed for knowledge binds the soul to material nature. Stanza 11 of this chapter indicates aura emitted by the body depending upon the dominance of one of the three virtues or GUN.

Sarva dvareshu dehesminprakash upjayate I
dnyanam yada tada vidyadvivrudham sattvamityut IIII II

(Gita 14, 11).

Light or aura (Prakash) from all the doors of body, is generated which is the light of knowledge and talent when SATVA GUN is dominant.

According to nature and tendency of the person which depends upon the dominance of the GUN the aura of the person changes. This is proved by KIRILIAN photography by modern science. This Clair science of auric expression was known to vedik civilization. Human tendencies are perceptible through exhibited auras by the body. Scientifically the tendencies are due to the quality particles in DNA of the individual. These particles get arranged according to the tendencies acquired by the prominent GUN out of the three gun or modes or virtues. Constant cultivations on inner core or CHITTA – mind-set of the individual. YOGI with pious SATVA GUN dominance and constant cultivations on CHITTA can get these quality particles rearranged and hence their aura gets changed. Different aura described in different persons:

1. Moon like bright VARN, aura is SATVA GUN dominant. This is seen in BRAHMINS. This aura is of good tendencies.
2. Reddish aura, VARN is RAJO GUN dominant. This is seen in warriors. These persons oppose injustice and oppressions are KSHATRIYA or warriors.
3. Yellowish aura VARN, is of VAISHYA. Such persons are alert to their own profit and benevolent to society.
4. Blackish aura, VARN is TAMO GUN dominant. Such persons are selfish, ignorant and belong to SHUDRA VARN.

Material world, which is in fact known as BRAHMA, is composed of three virtues/properties. These are SATVA, RAJ and TAMA. This terminology is scientific and holds science behind it. The evolutionary process which leads to truth and original nature is SATO GUN. The relative proportion of the three modes varies for every individual, due their *sanskars* (tendencies) from countless past lives, and accordingly, everyone has different inclinations and tendencies. Nature or Prakriti, is possible due to these virtues of material world. The supreme existence is

beyond the three virtues or GUNAS. Therefore in 2nd chapter Verse 45, Lord Krishna informs Arjun to be above these 3 modes.

Treigunyavishaya veda nistreigunyo bhavarjun II 2 II

(Gita 2, 45)

When spirit, comes in contact, with nature it gets bounded with these virtues. It is compelled to work within the effect of these virtues. Since different shapes, size and forms of life the spirit must follow the same nature because of bondage and effect of three GUNA – virtues, spirit is stimulated to perform according to species where existing. This is the cause of joy and grief for the entrapped soul. Analytical explanation goes further to state that as per work and effect/results of work the soul gets entangled into birth and death cycle again and again. But with determination of mind and performing assigned duty as per incarnation in any shape or form or creed, species etc. and renunciation of results of work performed to super soul, which is a privilege for human incarnation, one can attempt to free from such entanglement in cycle of birth and death. This is explained in GITA repeatedly. These three virtues are explained further and also, elsewhere, in VED, PURAN, PATANJAL yoga SUTRA etc.

Human personalities are developed from these three virtues. Person behaves, acts, or works according to the dominance of particular virtue. Explained below:

SATO GUN – PIOUS virtue or **mode of goodness**: This is best among all three because of it being pure and capable of providing knowledge, able to shed off sin. Those who have attained and acquired this virtue bound to get happiness and knowledge. Material miseries/pain/grief do not bother to such persons who have attained or acquired the SATO GUN. They progress in gaining knowledge and happiness. Bonded life of entrapped SOUL in material body, and so attracted towards work and continue doing work. Therefore, freedom from cycle of birth and death becomes negligible. They repeatedly undergo incarnation and become scientists or philosopher etc.

RAJO GUN – Kingly virtues – (**mode of passion**) Physical enjoyment predominate attaining this virtue. Opposite sex attraction and desirous of sense satisfaction, fame, social recognition, happy family etc. is outcome of this virtue. Indulges in hard work to attain above said satisfaction.

TAMO GUN – Ignorance, knowledge of darkness (**mode of darkness**). Stupidity, delusion, madness, sleep all these bound the possessor to materialism and material body. No advancement but degradation is the outcome of possession of such virtue. Ignorance generated TAMO GUN bind all living beings having material body, in a vicious processe Of MAYA.

Those who are involved in material body, are influenced by one of these GUN or virtues. When one is free from these nature's virtues and material sense he is free from bondages of MAYA and can be considered as liberated. In fact, a person has all the three modes of virtues in him. His behavior depends upon the dominance of one of these three. That decides the nature of that person. Hence Lord Krishna advocates Arjun to be above these three virtues

These are the answers to questions raised by ARJUN. In fact, VEDIK literature is full of concepts for liberation from "birth – life" cycle of material world. This is to have a better and fulfillment of life in a positive manner with perpetual happiness. Free from grief and sorrows of life. Truthful state of mind, True happiness what is called SAT-CHIT-ANAND. If one is involved and cognizant of material body TRI GUN influence state of mind. It forces body for sense satisfaction. At the same time when one fixes his mind and state of mind in SUPER SOUL, desire for sense satisfaction stops, Material body works as per nature's laws and life continues but free from involvement. Soul or KSHETRADNYA remains free. This position of soul is DIVYA – superior most position. Soul achieves freedom from TRIGUN influence. Such a liberated soul reaches BRAHM PAD (ब्रम्ह पद) – i.e. virtuously he attains similarity to BRAHMA. (GITA ch. 14, Stanzas 7-10). The state of person depends upon the dominance of the mode in that person. Similarly if a person dies, the reincarnation depends upon in which dominant mode the person dies at the time of death. Increased mode of ignorance, the birth is in animal kingdom. When increased dominance of mode is passion the birth is in among those engaged in works, with desire of fruits of the work. When death is in dominant mode of goodness, birth takes place in company of great sages (GITA 14, 14-16). **In other word, the resultant of mode decides the development of knowledge. With mode of goodness real knowledge develops, with mode of passion the greed develops, and with mode of ignorance the madness, foolishness, and illusion develop.** GITA 18th stanza says:

Urdhvam gachhanti satvastha mdhye tishthanti rajasah I
jaghanyagunavruttistha adho gachhanti tamasa II 18 II
(Gita Ch. 14)

Interestingly Gita has repeatedly reiterates the concept of re-incarnation as in Chapter 2, stanza 27:

Jatasya hi dhruvo mrutyudhruvam janma mrutasya cha I
tasmadapariharyerthe na tvam shochiturmarhasi II 27 II

Who so ever is born has to die is law of nature, after death, it is definite that rebirth too is part of the nature. Therefore, in performing duties destined to the person, there is no need to grieve on it.

Those situated in goodness mode, gradually go upwards to higher planes while in passion mode they remain in earthly planes. Those who live in ignorance mode, gradually go to worlds of hell. Shri KRISHN further clarifies in GITA, that any person belonging to any class, cast or VARNA out of CHATUR VARNA as described earlier, can achieve and practice the desired mode, can rise or fall according to constitution of nature. A BRAHMIN may fall down or SHUDRA can rise up as per mode the person involved in the mode of nature. They are interchangeable. Every person has right to change. It depends upon his attempt to rise or fall. This is the law of constitution of nature.

Social Science was highly developed in Sanskrit speaking civilization. The law of constitution of nature was recognized. Consideration for Goodness, Passion, Ignorance, i.e. different natures of all incumbents of society, so that tranquility, peace is maintained in the society. This ensures smooth and proper working conditions in the society. This is essential for all round progress of the human society. Greed, enmity and selfishness blocks the progress of humans. This is a big lesson to modern human societies of the world. The unrest is hinderance to progress. Proper discipline favors all round progress. Because of the proper discipline the Sanskrit speaking civilization achieved success in all the disciplines of science. In fact, it will not be out of place to say they achieved much higher planes of scientific progress in many branches of sciences.

Even though the Gita question-answers are about particular persons (Arjun and Lord Krishn), the idea and the conceptions are for the whole humanity and desirable for all.

CHAPTER FIFTEEN

PURUSHOTTAM YOG
ROOT OF UNIVERSE – SUPREME SOUL – CONTROLLER OF UNIVERSE
SCIENCE OF REGULATION OF BRAIN AND BODY

Abstract: Here in this chapter the material universe and manifest world is compared with an eternal tree. that originated on the Indian subcontinent, The tree is known as ASHVATH tree. It is a sacred tree native to the Indian subcontinent It is also known as the bodhi tree, pippala tree, or ashwattha tree (in India and Nepal). The sacred, fig is considered to have a scientific significance. Vedik meaning of ASHWATH is that it is a big tree and occupies big space. It occupies any where and everywhere like the soul omniscient and omnipresent. Grows anywhere. It releases oxygen day and night. The simile is given to manifest existence but with roots upwards and branches and leaves downwards. Our bodies are like this with control or roots upwards and rest body downwards. Multiple branches the network of nerves spreading downwards. Brain on top upwards which controls whole body which is downwards.

The supreme soul the eternal soul is upward and controls universe which is down wards.

Lord Krishn reiterates to Arjun that now whole higher knowledge be re-narrated by which the MUNIs have attained all the SIDDHI:

Param bhuyah pravakshami dnyananam dnyanamuttmam I

yajdnatva munayah sarve param sidhimito gatah II 1 II

(Gita 15. 1).

Excellent knowledge by which the MUNI/SAGES could attain SIDDHIs or perfection. What is science in this stanza? What are SIDDHIs or perfection?

Analytical description is available in Sage KAPIL muni's SANKHYA Darshan (Kapilmuni praneet tatvasamasah). 15th stanza says: "Ashtya sidhi" II 15 II. There are 8 types of SIDDHIs. (Stanza from SANKHYA KARINI):

Uhah shabdodhyayanam dukhavidhatastrayah suhritpraptihi I

danam cha sidhyotou siddhe purvongkushastrividhah II 59 II

Different types of perfections (Ashtasiddhi) in above stanza are:

OOH SIDDHI (perfection about previous birth and knowledge about elements): With sacramental knowledge of previous birth and studying Nature (PRAKRITI) acquiring knowledge about 24 elements.

Word Siddhi (knowledge about words): Gain knowledge from Preaching or instructions of Discerning and rational teacher.

Perusal siddhi (perfection in studies): Knowledge by studying and reading with scrutiny (Ved and Puran or SHASTRA).

Realized soul Siddhi (perfection in gaining knowledge from persons with realized souls): Getting knowledge by coming in contact with Realized soul (person) moving around to get rid of public ignorance.

Donation Siddhi (perfection **in** sacrifices): Yogi devoted to self-realization without caring for needs of the body. Gita in chapter 17 stanza 20, 21, 22 has analysed details about donation (DAN-दान).

The following Siddhis are resultant of above 5 Siddhis:
Spiritual sorrow Siddhi (perfection in spiritualization): Getting rid of spiritual sorrows.

Epiphyseal or sorrows related to fate (perfection in Ephemeral or fate due to supreme soul control-DEIVI): Getting rid of fate related sorrows.

Physical sorrows (perfection in physical effects): Getting rid of physical sorrows.

It is interesting to note that Hanuman, devotee of Shri Ram, achieved perfectionin all these eight siddhis (Ashta siddhis)

In an attempt, to convince ARJUN, in 15th chapter Shri KRISHNA has cited many examples. The examples are discussed one by one **to sift science from these examples.**

Gamvishya cha bhutani dharayamyaham mojasa I
pushnami choushadhihi savah somo bhutva rasatmakah II 13 II

(GITA, 15, 13)

This stanza has many hidden scientific facts. The scientific meaning of this stanza is discussed below:

To understand this stanza: Lord Krishna says that, "I" enter the living beings (Gamvishya cha bhutani) and 'I' support them (dharayamyaham \), with my energy (mojasa – shakti – energy), nourish and protect by vegetation using inexhaustible nitrogen cycle (pushnami/) – explained in earlier chapters. This 'I' execute by becoming MOON (somo bhutva – MOON). The SUN is essential for food production and synthesis of Chlorophyl etc. so is Moon essential for production of juice of life elixir of life (SOM-ras). SOM means Moon. This is using nitrogen from atmosphere. The Nitrogen cycle has constant Nitrogen amount it neither reduces nor depletes nor becomes more at any time nor its action capacity, is ever affected. The RIGVED very clearly establishes this fact which is at par with modern science:

Pushna chakram na rishyati na koshovpadyate |
no asya vythate pavihi ||3||
(Rigved mandal 6, Sukta. - 54, 3).

Nitrogen cycle never reduces neither its store ever changes nor exhausts.

Rigved talks about unity of moon (som-सोम) and nitrogen (Pushna-पूष्ण):

Somah pavate janita matinam janite divo janita pruthivya |
janitagnerjanita suryasya janitendrasya janitot vishno ||
(Rigved mandal 9, Sukta 96, Mantra 5).

Som keeps everything pure. It creates heaven (Dyu lok-दिवो) and earth, Sun gives inspiration to all and creates Indra and Vishnu and gives intellect.

Author has only summarized the meaning of Gita stanza 15, 13. One can realize from this that what voluminous science is hidden in this stanza.

Explanation of TOPSY TERVY TREE in this chapter:

Urdhvamulamadhahshakhamshvatham prahurvyayam |
chhandansi yasya parnani yastam ved sa vedvit || I ||
(Gita 15, 1).

The original source of universe is neither BRAHM nor VISHNU, but as per VEDIK literature. It is supposed to be from super soul the PARAMSHIV – TEJAHPUNJ. The supreme source of everything is the ultimate super soul or PARMATMA (परमात्मा). That is the root of everything – material, physical or spiritual world. In this chapter it has been depicted by a topsy turvy tree. A banyan/peppal/Ashvath tree, which has its root on top and rest of the plant is downwards Scientifically speaking there is no proof for this imagination. It is assumed and said that an imperishable banyan/PIPPAL tree having root on top and branches downwards, the leaves are VEDIK hymns. In Rigved mandal 1, Sukta 164, Mantra 20:

Dwa I suparna I sayuja I sakhaya I samanam I vriksham I pari I
saswajate iti I tayoh I anyah I pippalam I swadu I atti I anashanan I
anyah I abhi I chakshiti II

I. 164. 20.

Similar statement is in KATOPNISHAD CHAPTER 2, VALLI 3, STANZA 1.

Urdhvamulovak shakh eshoshvatthah sanatanah I
tadev shukram tad brahm tadevamruta muchyate I
tasminlokah shritah sarve tadu natyeti kashan I
etadvei tat II 1 II

Meaning a tree having the roots on top, branches down is an eternal tree. That is pure energy and all abodes are situated in it. (see abstract).

In Sankhya yog, Rigved the story of two birds has been described. One wanders over branches and busy in material gains and material satiety the other aloof from such things and involved with supreme being. In other words, two living entities – Ishwar and Soul – remain together like friends always. Similarly, another entity (Nature) exists like a tree with branches. One of the living entities – the soul – tastes the fruits on the branches. The other entity – Ishwar – is completely away from this and hence never gets into worldly things.

One who knows the tree is a knower of VED. Involvement in this material world is explained by the banyan/Pippal tree. One who is deeply involved in sense satisfaction of this material world, wanders from one branch to another and one leaf to another and gets no satisfaction at the end. This is like MIRAJ in desert. There is no end to the sense satisfaction in material world. This is root cause of entanglement. Realization of this fact is in fact first step to achieve liberation. The tree "topsy turvy" – root upwards is symbolic and depicts a fact that roots of our existence in material world are descending from top – BRAHMA which is from center of VISHNU, having connection with SHIV. What is the hidden science in this stanza, is described below:-

Space sciece is involved in this statement. In astrology (JYOTISH SHASTRA) time calculations (Kaal ganana) is integral part. In Gita at many places such statements are narrated as:

Kaalah kaalayatamaham-(Gita 10, 30) – "I am Time among countable time".

Jyotisham raviramshumana (Gita 10, 21) – Time depends upon SUN. "my" form is SUN. There are 27 NAKSHATRA. 2 ¼ Nakshatra is one sign of Zodiac (Raashi – in astrology it is moon based and not sun). There are 12 Rashis and sun moves in these Rashis i.e. sun is for 1 month in each rashi. नक्षत्राणामहम् शशी (10, 21) – "I am moon among nakshatra". Every 6 months is AYAN i.e. two AYAN in one year. Two SOLAR SOLTICEs in one year – UTTARAYAN and DAKSHINAYAN:

Agnijyotirah Shukla shanmasa ottarayanam I
tatra prayata gachhanti brahma brahma vido janah II 24 II
(Gita 8, 24)

Ghumo ratristatha Krishna shanmasa dakshinayanam I
tatra chandramasam jyotiryogi prapya nivartate II 25 II
(Gita 8, 25).

Other calculations already explained earlier. Ultimately ETERNAL TIME or AKSHAY KAAL is supreme soul or Param aatma, so Lord Krishn says –

Ahamevakshayah kaalah II 33 II
(Gita 10, 33).

In other words, the time ceases to exist in ultimate supreme soul. Hence the supreme soul is beyond the time and described as KAALTRAYATEETAH (in Yajurved, Atharvasheersha).

It would not be out of the way to mention here that the ancient SANSKRIT speaking civilization developed expertise in ASTROLOGY, PALMISTRY and ASTRONOMY etc. The art and science of planetary positions, planetary rotation in space, different constellations (NAKSHTRA-27 in number) their effects on different endocrine glands, tissues and organs (already described) the reasons of diseases presented in hospitals in groups and crops majority of times and not alone or sporadic cases. 27 NAKSHTRA influence nuclei of brain and in turn the whole body. Nature or super soul is present everywhere and pervades each, and every particle of nature smaller than smallest or bigger than biggest. The Laws of nature are within the constitution of universe. Manifest and un-manifest world is governed by these laws (already mentioned in introductory pages of this book). The law, structured in the DNA of every cell is in totality, the same natural law and law of automation. **The whole physiology is governed by the automation principle of expanding universe and the living body in the material world**. The relation of individual with cosmos, or one can say that, influence of cosmos is on every grain of creation in this universe. In fact, the supreme soul organizes the whole universe by its natural laws. In gita stanza 7 it is clearly emphasized that:

Mameivansho jeeva loke jeeva bhutah sanatanah I
manah shashthanindriyani prakrutitasthani karsjati II 7 II
(Gita 15. 7.)

All living beings are fraction of eternal super soul. Life lives in physical existence, with the help of mind and Senses. How mind and sense are governed? To explain scientifically the automation governance of super soul following explanation is important.

To elaborate this Maharishi Mahesh Yogi's Vedik Science and technology is very useful.

There is direct relation, direct action, and influence of SOLAR SYSTEM on brain. There are 9 planets in solar system. These 9 planets influence the 9 elements of brain:

⌖ SURYA – Sun has influence on Thalamus

⌖ CHANDRA – moon has influence on Hypothalamus,

⌖ MANGAL – Mars has influence on Red Nucleus, Amygdala

⌖ BUDH –mercury has influence on Subthalamus

⌖ GURU – Jupiter has influence on Globus Pallidus–

⌖ SHUKRA – Venus has influence on Substantia Nigra (mid brain)

⌖ SHANI – Saturn has influence on Putamen – (Regulate movement and learning)

⌖ RAHU – SHEWET – Uranus? Has influence on Nucleus caudatus – Head

⌖ KETU – Neptune? – Has influence on Nucleus Caudatus – tail

(Uranus and Neptune names of Rahu and Ketu are debatable in English, and hence the question mark)

Similarly, Solar system influences DNA constituents – Guanine (GURU, Sign of zodiac, Capricorn or MAKARA), Adenine (SHANI – KARK, Cancer), Cytosine (MANGAL, – TULA – Libra), Thymine (SHUKRA Sign of Zodiac – MESH/ARIES), Sugar (BUDH), Phosphate (CHANDRA), Enzymes (RAHU, KETU). 12 parts of Nucleotides of DNA have their counter parts in 12 RASHIS – signs of zodiac.

Through – Planets (GRAHA) Super soul organizes, directs and controls (through solar system).

Human brain is influenced by universe. The 12 signs of zodiac – the RASHIS, and cortical areas of brain have one to one relationship. As per Horoscope chart or RASHI KUNDALI have 12 houses (BHAV). Each house has its own influence on the individual. These BHAV (house in a kundali-map of celestial satellites – graha) are influenced by the 12 RASHIS which in turn influence the cortical areas of brain.

1ˢᵗ – TANU BHAV – Generally is about birth or "Janm Bhav". Gives idea about what a person gets at birth. In brain it is **Right and Left Occipito-temporal area – self-image memory, Identification – Aries.**

2nd – DHAN BHAV – It forecasts persons Prestige and wealth. **In Brain it is Right Occipital area – Vision, face identification, appreciation of objects – Taurus.**

3rd – SAHAJ BHAV – It identifies courage, travel, research, art etc. of the person. **Right Parietal area sensory representation of hand, arms, Skill, perception of space– Gemini.**

4th – BANDHU BHAV – It foretells Feeling and desire of the person. **Right Limbic area – Emotional, instinctive function, Gustatory and Olfactory association – Cancer**

5th – PUTRA BHAV – Persons Learning, knowledge, progeny is predicted. **Right Frontal area – Memory, learning – LEO**

6th – ARI BHAV – About enemity is indicated in this house. **Right Prefrontal area – mood regulation, motivation, Mental diseases – Virgo**

7th – YUVATI BHAV – Life partner, partnership, heroism etc. is indicated in this house. **Right-Left Temporal area –Sensorial function – Libra**

8th – RANDHRA BHAV – Age, is dealt in this bhav. **left pre-frontal ar**ea **– Scorpio**

9th – DHARM BHAV – Spiritualism, long life is considered in this bhav. **Left Frontal area – Reasoning, prognostication, interest in Occult – Sagittarius**

10th – KARMA BHAV – Karm and karm field is indicated. **Left Limbic area elaboration of personality, Vocation – MAKAR – Capricorn.**

11th – LABH BHAV – Opportunities predicted from this bhav. **Left Parietal area – Sensory integration, Physical comfort – Aquarius**

12th – VYAY BHAV – Family pleasures are indicated from this house. **Left Occipital area – Visual discrimination, Temptation – Pisces**

Through this arrangements celestial and planetary relations, super soul regulates and controls with automation. That is how Vedik statement "Super soul is charioteer: which becomes self explenetory (Chapter 10, Brahma bhavati sarathihi, (RIGVED, 1. 158. 6).

Gita stanza: Gamvishya cha bhutani dharamyaham mojasa, stated above, in the form of energy-Shakti or soul contained in all living beings, the protection, nourishment and sustain life with the help of MOON by providing *juice* of vegetation and medicinal plants taking help from inexhaustible Nitrogen cycle (PUSHNA CHAKRA), which is also under control of super soul. Similarly, same thing is highlighted in rigved. This is a very small statement in Rigved, but has a very videspread meaning. Super soul (SHIV) controls Primordial gasses – VISHNU. The primordial gaseous layer protects BRAHMA from harmful effects of radiation, light (TEJ), heat and many more unknown rays emitting from super soul-SHIV, (which are still unknown). Modern science needs more research in that direction. In other word, supreme authority of universe is in fact a charioteer for all living beings. In Chapter 10, it has been explained that how Brahm and Living beings are created. This is explained with the help of and citations from AITTARIYOPNISHAD. This Brahm under protection of Vishnu becomes charioteer. To explain it further:

SARATHI means driver of chariot – a horse driven cart. The cart is under control of driver with reigns to control horses and guide the chariot. Super soul after creation of universe and galaxies, stars, different constellations, groups of stars and planets like solar system, with the help of automation reigns over living beings. Cosmic rays from celestial elements of cosmic manifestations, using as reigns control the living beings of whole universe. The controlling mode has been explained above, with the help of Constellations, Nakshtra, Sun and Planets of solar system. Indirectly the super soul becomes Charioteer. This is the meaning of RIGVED stanza and Gita that super soul controls the whole universe.

Gita says here that super soul fractionalized enters the planets and with that energy they stay in orbit. Fractionalized super soul in moon controls activity, there by supply the juice of life all plants and vegetables, to energize all living beings of material world. Thus, they depend upon super soul and so "BRAHM BHAVATI SARATHI" or BRAHM is charioteer.

Accurate knowledge and understanding of space, stars, and planetary positions, of Sanskrit Speaking civilization, led to establishment of many scientific standards. Development of accurate calculations about celestial phenomenon including eclipses of moon and sun, even today with second to second accuracy of beginning and end of the celestial occurrences-the eclipse. This in fact is a science. Wonderful part is their effects on human life as per their zodiacal signs has also been derived and presented. This cannot happen without scientific studies and mathematical accurate calculations. **Accordingly, the tree description with roots up and branches, leaves downwards is for the understanding the indestructible, unperishable tree of illusion**. When one understands it the possibility of getting out of the tree would be easy. The material entanglement involved with such a tree can be shun off. This unreal fictitious tree can be cut by weapon of detachment (असंगशस्त्रेण). One can reach the place of supreme soul – a place from where all is manifest and to the same everything returns. One who understands the eternal, has given up material lust, not disturbed by grief – happiness, auspicious-inauspicious, loss-gain, success-failure and with steady state of mind (तटस्थचित्त) surrendered to supreme soul can attain eternal kingdom. A place having no sun, moon, fire and having reached there can never returns, back to mortal world. Such is the abode of supreme soul. Stanza 6 of chapter 15:

Na tad bhasayate suryo na shashanko na pavakah I
yad gatva na nivartante taddhamam paramam mam II 6 II

(Gita 15, 6).

Everything of material world manifests from this place and ultimately un-manifests at the time of annihilation. How then the Sun, Moon, Fire, or the material world can exist at such a place? The fragmented part of supreme soul in material body becomes spiritual body which is at par with supreme. The shine, light and heat of sun, moon, and fire are existing due to supreme, in fact created by supreme. In scientific description the whole universe – the galaxies, stars, constellations which are created by Brahma and their protector VISHNU and their material manifestation is due to supreme. Everything receives energy from that supreme. Everything is receiving energy for rotation of galaxies stars etc. from supreme. (see brahmotpatti)

The stanzas in Sanskrit are having deep hidden aspects of science but superficial meaning get camouflaged by religion. The science was a part of living of the Sanskrit speaking civilization. Each and every incumbent, of the population, was knowledgeable (vide Lord Macoulay's statement in British parliament in 19th century. Chapter 3 of this book page 103). The glorious sunshine of knowledge needs verification and support statement from rulers of enslaved country. This is effect of "TIME CYCLE".

The scientific theory of birth and rebirth is explained in following stanzas of Gita.

Mameivansho jeeva loke jeeva bhutah sanatanah I
manah shashthanindriyani prakrutitasthani karsjati II 7 II

(Gita 15. 7).

Shariram yadvapnoti sachchapytkamatishwarah I
Grihiveitani sanyati vayurgandhanivashat II8 II

(Gita 15, 8).

Entrapped soul in material body is fragment of super soul (mameivansho jeev). It controls mind and 5 senses, total 6 (manahshashthanindriyani). At death and birth, it carries all 6, as wind carries aroma with it (vayurgandhanivashat). The soul carries this between birth and death also before it is confined in new material body. The mind-intellect and 5 senses or the bodyless soul possesses scientific Clair knowledge, even after death. It can be derived that the soul with six elements (manahshashthanindriyani), exists as ethereal body (LING DEHA), in space after exiting from body after death. Clair science as per Yogik knowledge of SAGES other theory is JEEVATMA automatically get attracted towards the developed physical body in the womb. This is directed by the Clair vibrations from the foetal body in womb. The vibrations are at the highest pitch by 7th month. Similar vibrating pulsations emitted by the wandering soul. This is comparable to radio frequencies from radio station to radioset. Same opinion is in Shrimad Bhagvatam given below.

Shrotram chakshuh sparshanam cha rasanam ghranmevacha I

adhishthay manashchayamvishayanupasevate II 9 II

(Gita15, 9).

Soul enjoys the objective world, with senses – Eyes, ear, touch, tongue, nose etc. The mind enlivens the sensual world. Material physical body is an instrument of the soul for the function of JEEVATMA.

The step wise description of growth of foetus is at par with modern science in Shrimad Bhagwat by Shri Kapil Muni (ferilization of ovum with sperm, Kalal and then budbud formation and so on). The Clair science developed by YOGIK knowledge, about the growth of foetus in womb from conception till 7th month and at this stage the development of consciousness takes place with the entry of the wandering soul into the material physical body. But no scientific evidence could be traced by the author. Further description is still interesting:

The formed body till prior to 7th month is unable to survive independently. With entry of soul in 7th month the body becomes knowledgeable and conscious at the same time capacity to survival is developed. JEEV gets degree (PADVI) automatically. In other word the birth takes place at 7th month of the pregnancy and not on the day of fertilization of ovum with sperm. Therefore, the ASTROLOGICAL calculations are decided at birth and not on the day of conception. It is further stated that the developed body of foetus functions with the help of vital force of mother and not the soul or JEEVATMA. At the end of 7th month the JEEVATMA takes over the functions. Modern science is silent on this. But this is a subject of serious research for the modern scientists.

Aarabhya saptaman masat labdhya bodhopi vepit I
neikabhaste sutivateihshthibhuriva sodarah II10 II
(Shrimad Bhagwatam Skandha 3, chapter31, mantra10).

Mantra10, Thus, endowed with the development of consciousness from the seventh month after his conception, the child is tossed downward by the airs that press the embryo during the weeks preceding delivery. Like the worms born of the same filthy abdominal cavity, he cannot remain in one place.

Tam jeevkarma pdvim anuvartamanah taptrayopshamanay
vayam bhajet II 16 II.
(Shrimad Bhagwatam 3, 31, 16).

Mantra 16: No one other than the Supreme Personality of Godhead, as the localized Paramātmā, the partial representation of the Lord, is directing all inanimate and animate objects. He is present in the three phases of time — past, present and future. Therefore, the conditioned soul is engaged in different activities by His direction, and in order to get free from the threefold miseries of this conditional life, we have to surrender unto HIM only.

Scientific explanation is as per the quality particles of DNA already explained above.

(BHAGWAT GITA – Vyas meaning, by Yogi Manohar. Published by Shri Trimbakrao Manohar Harkare and Shri Dattatrey Manohar harkare, Tulsibag road Mahal. Nagpur-440002).

It is clear from above statement and facts that birth of soul is under direct control of Super soul. Scientifically it depends upon the frequency matching of soul and the body in womb. In short, the unevolved body can not accept the evolved soul and vice versa. Evolution of soul depends upon the good cultivation in the body. It is in the hands of the person and the peoples with wisdom to inculcate the good cultivations in the society which has the capacity to even change the quality particles in genes and chromosomes in the cells of the body.

Understanding this, the Vedik culture was developed by the then civilization. Which resulted in all round progress socially, scientifically and spiritually. This theory is for all the global civilizations today.

CHAPTER SIXTEEN

DEVASUR SAMPAD VIBHAG YOG
LIVING BEINGS AND INDIVIDUAL NATURE
DIVINE AND DEMONIC NATURE
HUMAN PSYCHOLOGY

Abstract: Demoniac and divine nature of population on which depend the bad or good behaviour of the people is discussed in this chapter by lord Krishn. The tendencies of nature are explained. Hence symptoms of such behaviour are narrated. Scientific aspect of such tendencies are explained. Interestingly this narration is description without any questioning by Arjun.

The opening stanza of 16th chapter states the way of behaviour of each and every person, of the society no matter to which VARNASHRAM system he belongs. (this has been discussed in previous chapter – as per VEDIK norms) This (VARN or ASHRAM) is for the welfare and peace in society, community, and the countries of the world. Because of such mind set VEDA preaching is for VASUDHEV KUTUMBAKAM. This quality OF HUMAN behaviour is more relevant in modern time. In first three stanzas good qualities in godly humans are clearly narrated.

In Gita chapter 4, stanza 13 Lord Krishna says that:

Chaturvarnyam mayasrushtam gunakarmavibhagashah I
tasya kartaramapi mam vidhayakartaramvyayamII 13 II
(Gita 4. 13).

"Even though 'I' have created this system still 'I' am nonactive and undutiful in this CHATURVARNA system".

Apart from divine and spiritual qualities (detailed description of qualities is given below. Reader is advised to consult original Gita book) of a person mentioned in stanza 1 and 2 of 16th chapter, Gita speaks about how to generate progeny with divine qualities with the use of GARBHADAAN Sanskar. Stanza 3 refer to this in the bold word "DAIVIMABHIJATASYA" (below):

Tejah Kshama dhritihi shouchamdroho natimanita I
bhavanti sampadam deivimabhijatasya bharat II 13 II

Vigor, forgiveness, fortitude (courage in adverse conditions), cleanliness, freedom from envy, and no desire for honour are genetically bestowed qualities, found in a person. All these qualities form transcendental nature of the person. The pious qualities, as per previous chapter, are also from previous birth.

There are 16 SANSKAR of which one is GARBHADAN sanskar in Vedik Sanskrit literature. It deals with procurement of progeny with divine qualities. ABHIJATASYA word has hidden scientific meaning. It is an understanding that Gita advocates restrain from sex life (MAITHUN). On the contrary procurement of progeny must be in a scientific manner. Qualities of progeny described in literature which make society pious and auspicious are mentioned in first three stanzas of GITA 16, 1, 2, and 3. Transcendental qualities mentioned in GITA are for better and peaceful, prosperous, progressive society, like – Fearless, pious, pure and Spiritual, knowledgeable and follower of knowledge), Tendency and propensity for donation – Charity, Self-control, expert in work, studious (in Vedik knowledge), Penance and austerity, Simplicity, nonviolence, Truthful, Anger less, Renunciation, Peaceful, uninclined for fault finding, compassionate in all living beings, ungreedy, Modest, Determined, Full of Valour, Forgive full, Unenvious. Certainly, such qualities in persons of society will make country auspicious.

In TAITARIYA Upnishad (2-Brahmanandvalli), 8 (ANUVAK), 2. it says:

Shanandasya mimansa bhavati I
yuva syatsadhu yuvadhyakah I
ashishtho dradishtho balishthah I
tasyeyam pruthvi sarva vttasya purna syat I
sa eko manusha aanandah II 2 II

On serious thinking about pleasure (Meemansa), if a person is young educated, courageous, strong willed, educated, strong then he can acquire the whole earth, such a pleasure is considered a unit of human pleasure. Hence such a progeny is desire of every couple.

Sanskrit literature on the subject of scientific procurement, of progeny can be discussed in following groups: (see stanza of Manusmriti 3/5, 12, 42. below)

1. Before marriage selection of bride and bridegroom. Gotra is genetic lineage of father of bride and bridegroom should not be same. Such marriage is SHUBH or pious, auspicious and favorable for progeny procurement. This leads to abnormal off springs. **As per modern science the DNA incompatibility can be avoided by Gotra matching.** It needs more research on the subject.

2. Love marriages are generally with physical attraction and for pleasure, may not be auspicious or scientific and progeny may be inauspicious. Targeted research is essential as per modern science for making better country men.

Asapinda cha ya matursagotra sha ya pituh I
sa prshasta dvijatinam darkarmni meithune II

aavarnagre dvijatinam prashasta darkarmani I
anindateihi strivivahernindaya bhavati praha I
ninditeirnindita nrunam tasmanindayan vivrjate II

(Manusmriti 3/5, 12, 42).

In Sanskrit literature and Vedik opinion empowerment and wholeness of a society is considered when Health, Education, Confidence (undisturbed in difficulties), courage, Property, and Enjoyment are six equally taken care of. Taittariyopnishad as above.

3.   Certain days (TITHI) are to be avoided for Garbhaadhan. During Eclipses, Amavasya (30th day of lunar month), Ashtami (8th day of lunar month), Pornima (15th day or day of Full moon), and Chturdashi (14th day of lunar month) Garbhadhan be avoided. Garbhadhan on such days leads to inauspicious progeny. (Manusmriti 4/128)

4.   Bharatiya Sanskriti (Vedik Civilization) has considered GARBHADHAN as Scientific and spiritual, to produce the ideal population for the society and the country at large. It is evident from the overall progress and prosperity to the extent invaders considered BHARATVARSH as GOLDEN BIRD. Invasion was with the idea of looting the wealth.

In Vedik literature all statements are, to drive in the idea of procurement of ideal population as per Vedik standards.

5.   Institution of ideal marriage, (SHUBH VIBAH), was based on scientific basis. Garbhadhan was not for sense gratification or greed in enjoyment only. Garbhadhan is placement of fetus in the form of seed in womb of life partner. This is achieved by following above said principles and avoiding inauspicious days and time etc.

(BRIHDARANYAK Upnishad, Adhyay 6, Brahman 4, SHLOK 21; and PRASHNOPNISHAD Prashna 1, shlok 13).

VEDIK way, of order of life, and as per MANUSMRITI description of society behaviour VARNASHRAM DHARMA as well as ashram way of life order has been misunderstood and misinterpreted in current social behaviour under the influence of invaders who ruled Bharat. These rulers were having vested interest. It (VARN or ASHRAM) has been labelled as cast system or cast society. The division of incumbents of society was not according to birth but on the basis of work and intellectually oriented system, mentality and educational level of achievements. It was not having any intention to divide human society on the contrary, it was to unite and live with harmony and brotherhood. One can imagine that if the preaching is for "the world is one family", then how incumbent of the society be divided? As mentioned earlier in modern world and in all the countries the "Varnashram system" is existing based on work and intellectual aptitude. Incumbents of society are grouped as per their work performance for example bureaucrat, Technicians, scientists, doctors and businessmen etc. This is **relevant as work perfection is possible as per expertise of person**. This is as per every persons' DHARM. Same is in VARNASHRAM. Not only this, a person according to his age and wisdom with time in life capability of individual alters according to stage of span of his life. Peace and tranquillity in society helped to extraordinary progress, prosperity and scientific achievements. ASHRAM system was prevalent during VEDIK period of Sanskrit speaking civilization. Groups described are BRAHMACHRYASHRAM, GRIHSTASHRAM, VANPRASTHASHRAM or SANYASASHRAM. This is for the steady progression of every person of the society. At the same time four aims of life DHARM, ARTH, KAAM and MOKSHA can be easily fulfilled. Religion, economy, sex and liberation must be enjoyed by each and all. Living like animals for sense satisfaction is not the only function of life for humans and brainy people. Liberation is also important and must be considered in life. For this, practice of austerity, simplicity, renunciation, tranquillity, forgiveness, nonviolence, spiritual knowledge, charity, self-control, sacrifice and study of VED is advocated by Shri KRISHNA himself in this stanza. By following such principles, one becomes simple and straight forward.

Dambho darpobhimanashcha krodhah parushyameva cha I
adnyanam chabhijatasya parth sampadamasurim II 4 II

(GITA 16, 4).

In stanza 4, the qualities of a person are described which takes a person to further bondage to material world. The liberation in fact is forfeited. Pride (dambh), arrogance (darpR), Conceit (abhiman), anger (krodh), harshness (parushya), and ignorance (adnyan) etc. are all qualities of demonic nature (asuri pravrutti). This nature of humans prevents him from the aim of life, of liberation. If all humans inculcate such nature what will happen to the world society? Such inauspicious nature will lead to unrest, war. Peace of the human society will be disturbed resulting in halt of progress and prosperity. Whole world will become unliveable and population will perish. Transcendental nature is always conducive for liberation and demonic nature lead to bondage and greed for sense satisfaction. Demonic natured person fails to understand the proper or improper behaviour, cleanliness of body or mind. Truthfulness is missing in such a nature. Such natured mind is lost, in its own wrath. Unbeneficial and horrible nature is bound to destroy the world (kshayay jagatohitah). Demonic minds engage in cruelty to animals and humans as well. Development of nuclear weapons (prabhavantya ugrakrmanah) creating pride feeling amongst the different countries use of which is bound to destroy the world. In what way such inventions are for the prosperity or development and progress of human population of the world? Such demonic nature always feel and believe, in gratification of the senses, greed to satisfy own mind-set. It is the ultimate aim of life, for such natured persons. All the desires are full of lust and greed. **Money procurement is by unfair illegal means (kaam, bhog artham anyayen arth sanchayam\). (GITA Chapter 16. 11 & 12).** All such thing are delusion of ignorance and ultimately dwell in hell (patanti narakashuchou).

What compels Lord KRISHN to narrate demonic and pious nature to ARJUN? A simple reason seems to be to balance after understanding comparative statement of nature ARJUN can decide to adopt the nature which is pious and dutiful according to his DHARMA, for himself and the population at large. Duty performed piously keeps a person free from results of work done, good or bad (GITA 2nd chapter).

Thus, after explaining good and bad aspects of virtuous and demonic nature, Shri KRISHN leaves ARJUN to decide.

Here in 16th chapter Krishna has given examples of Genetics (abhijatasya) to explain to Arjun.

How genetic science plays part in this pious or demonic nature? Even if one is engrossed by genetics and has nature accordingly, the person is at free will to transform to betterment or to detrimental nature by his own decision and mind-set. There is no genetic binding of any kind. State of the mind can decide the upward or downward journey of embodied soul. At this juncture the role of SANSKAR is evident. Cultivations the "Sanskar" can be good or bad. This depends upon family traditions, teaching and the society. If demonic body entrapped soul decides and alters his state of mind for self-realization and pious behaviour, then upward journey and uplift of entrapped soul is a possibility and vice a versa. The soul as per laws of nature indulging into greed, lust, anger etc. gets another body in life's species for degradation of soul. This indicates two things. One is genetics while other is rebirth or reincarnation. Which "species body" the person will get in next birth or reincarnation, is not in the hands of the individual but super soul decides according to his KARM PHAL (WORK RESULTS). Modern science is still investigating the possibility of rebirth theory. Near death experience in different humans getting solid ground in this direction. Many real – life stories witnessed by investigating scientists, forces modern scientists to believe the theory of re-incarnation.

Jatasya hi dhruvo mrutyurdhruvam janma mrutasya cha I
tasmadpriharyerthe na tvam shochitumarhasi II 27 II
(GITA 2/27 stanza).

One who is born, must die, and after death is sure to take birth again, (rebirth). This cycle goes on repeatedly till liberation. It is work result oriented. Unnecessary murder, slaughter, crime, war, are inevitable in human society. Fight for the right cause is duty of a KSHATRIYA – the warrior. Hence no one should lament from discharge of duty. Such an act will certainly lead to liberation. Therefore, there is no need to lament on discharging duty.

Na hi kashchit kshanamapi jatu tishthtyakarmakrut I
karyate hyavashyah karma sarvah prakruti jeirguneihi II 5 II
(GITA 3/5 stanza).

No human can be without thought or work even for a second. One is constrained to perform duties as per qualities acquired by modes of nature.

All varieties of Works generate result. Result can be immediate or delayed. As per VEDIK philosophy KARMPHALA (result of work) is seen in the same life span or some can be seen in other – next incarnation. Pious work results into upliftment of entrapped soul and demonic work repeatedly leads to birth and reincarnations. After self-realization only one gets out of this cycle of rebirth and reach the PARAMPAD – the goal of life.

Tamev viditvati mrutyumeti nanya pantha vidyateyanay II 8 II
(Shwetashvatar Upanishad 3, 8).

After the true knowledge of Super soul (PARAM AATMA) only, one can break the cycle of birth and rebirth.
(Details in – PARLOK AUR PUNARJANM KI SATYA GHATNAYE – GITA press Gorakhpur, 273005., ISBN 81-293-0276-4).

Scientifically how this rebirth is possible? How the memory of previous birth is retained? What is the role of soul which is a fraction of super soul or energy from the source of ultimate energy the super – soul? As mentioned earlier, GITA is only a question and answer form between ARJUN and Shri Krishna. Only point of view is endorsed with examples of knowledge existing in society and population, which are well established and prevalent in the society. Details of embryology are described in Devi MADBHAGWAT PURAN a part of SHRI MADBHAGWAT.

Development of cell mass after fertilization has been named as KALAL, BUDBUD etc. = This is narrated to father by a daughter, who is supposed to be incarnation of Goddess. This is to explain how one can achieve liberation from life cycle. The cells have brain and memory. This can be explained by the use of cells, for regeneration of tissues and organ in vivo and in the laboratory (Authors own research). When Ovum is fertilized with sperm it acquires a tremendous capability to divide and re divide and form tissues and organs having different structure and different functions. This only is possible when cells retain the memory to form those tissues and organs. With that memory the cells exhibit their capability to form tissues and organs desired in the region.

(Matapurkar, B.G. A New Physiological phenomenon of mammalian body for organ and tissue neo-regeneration in-vivo: Adult stem cell technology in the perspective of literature. Indian J. of Experimental Biology. Vol 40, 2002, page 1331-1343.)

Is it that the cells retain memory in rebirth or new embodiment in material world? But in rebirth the memory of different site and different families not connected to each other have been described. It is observed that the memory is expressed in context with different family and different sites far away from the family of birth of the concerned individual. In such situations genetic connection is absent then how the memory is carried forward?

In a stanza from AITTAREYOPNISHAD, Chapter 2, section 4, stanza 1 to 4.

Purusha ha va ayamadito garbho bhavati tadetadretah yadetsarvebhyomgemyastejah sanbhutamatmanyevatman bibharti I
tadyada striyam sinchatyatheinjjanayati I
tadasya prathamam janma II 1 II

(AITTAREYOPNISHAD, Chapter 2, section 4, stanza 1 to 4).

Soul energy (purush) in the beginning (ayamadito) present inside (garbh) the male in semen (tadetadretah = ret), which is a product of all energy (TEJ) of body (sarvebhyoum gemyastejah) which self-nourishes itself (sambhutatmanyevatmanam), needs no outside help, when sprayed in female (sinchatya) it is born. **This is first birth.** Three scientific facts are noted in this stanza (1). Soul energy first in male semen which is Squeezed extract – TEJ – of whole body. After ejaculation body is energy less and weak. (2). Self-nourishing semen, the sperm needs no nourishment from any outside source. It is self-nourishing. Modern science agrees to it. (3). Birth starts when in female the Soul takes its first birth. Another fact about whole body energy is, it produces another new body. Because modern science knows that the sperm has coded language and memory stored in sperm. Without any support it can regenerate other human. It grows in the female body, but it is not only independent, it transforms body of female to its advantage. The process is totally dependent on automation. Best example for this is when an artificially fertilized ovum kept in a 60 year old female uterus, (when reproductive capability of female is at standstill or finished), it converts the body of that female to its own advantage for help and support till full term of pregnancy. The post-delivery, the nourishment, is ensured by developing breasts for lactation in pregnant female.

It is surprising that such a knowledge about sperm, semen, its nourishment etc.; was prevalent with Sanskrit speaking civilization about 5000 years BC.

Sperm needs no support, when the growing foetus in uterus is wholly dependent for nourishment on mother. The sperm is not connected to any blood circulation or drainage. It is not depending on body of the possessor-father. Other fact is it can survive in female body for 5-7 days also till ovum is fertilized. This justifies the Sanskrit statement of Rishi Aittareya. Second stanza is still astonishing to modern scientists.

Tastriya aatmabhuyam gachchhati yatha svagamam tatha I
tasmade nam na hinasti sasyeitama tmanmatra gatam bhovyati II2II

The female body considers semen (RET) as her own body (swamagam) hence no rejection process (na hinasti), Blood of husband and wife may not be same or matching, is rejected but semen is not. Same is with transplant surgery between husband and wife, unmatched organ is rejected. Third stanza says:

Sa bhavyitri bhavyitavya bhavati tvam stri garbham bibharti sogra eva kumaram janmano gredhi bhavyati I
sa yatkumaram janmanogagre dhibhavayatyatmanmev tadbhavayatyesham lokanam santatya evam santata hime lokastadasya dvitiyam janma II 3 II

The female protects, cultivates, grows, nourishes the foetus, in her body (Garbham bibharti) and even after birth, the female continues this. That is how the population continues. This is soul's **second birth**.

Sosyayamatma punyebhyah karmabhyah pratidhiyate I
athasyayamitar aatma krutakrutyo vayogatah preiti I
sa itah prayannev punarajayate I tadasya tritiyam janma II 4 II

It is stated that:

Soul has **three births** from preconception to delivery of child.

Is it that the soul is responsible to carry the memory, which is capable of moving far and away. This is a point of research for the future investigators on the subject, of rebirth, or reincarnation. In this context it is mentioned here that the foetal development takes place till 9th month in uterus. Now at this point the total CHAITANYA SHAKTI (life energy) received by the developed foetus.

Navame mase jivastu cheitanyam sarvsho labhet II 28 II

Matru bhuktanusaren vardhate jathare sthitah I
prapya vei yatana ghoram khidyate cha swakarmatah II 29 II

(DEVI PURAN (mahabhagwat) SHAKTI PITHAK. NO 1, YEAR 79, PAGE 159.)

At this stage and during delivery extreme distress and agony is experienced by the individual, accordingly to his KARM PHAL – work done and its results in previous life, and feels extreme disgust.

He – the jeev – decides to follow super soul – the PARAM AATMA AND ASSURES HIMSELF FOR PIOUS BEHAVIOUR IN LIFE TO FOLLOW. After birth, the effect of camouflaged nature (MAYA) he forgets everything later and starts enjoying life. Sense gratification and its desire keeps him into birth death cycle again and again. The material body remains involved in sense satisfaction, the soul has nothing to do with this sense satisfaction. It is separate than the body. But with efforts (PURUSHARTH) this can be effectively changed.

Sa dehah purushadbhinnah prushah kim samashnute I
pratikshanam ksharatyayushchlatprnasthatoyavat II 43 II

(DEVI PURAN (mahabhagwat) SHAKTI PITHAK. NO 1, YEAR 79, PAGE 159.)

The self is separate than the body which has hardly any connection with the body enjoyment but is silent spectator of actions and its results (Karma phal). The stanza 29 above emphasizes a fact that even in intra uterine life "the self" realizes past incarnation which is KARM PHAL EFFECT. But how the past incarnation memory is retained is topic for research. This needs research on intrauterine brain activity with the help of investigations using available modern science instruments.

It can be reiterated here that, all living beings are afraid of death. Where from this fear has come and why? No one has any death knowledge or experience in present incarnation. But still a boy of two years if peeps from height of tall building, he with reflex steps back due to fear when he has no previous experience. Somewhere, experience of previous incarnation is retained which warns the living creature? This probably help to accept the reincarnation theory. There are certain facts which cannot be directly proved but indirect evidence becomes satisfactory. Electricity current, magnetism, wind is acceptable by indirect evidence.

Refer Gita 15, 9, where in it is clearly indicated that the soul carries with it the 6 (5 senses and mind), elements at the time of death from previous body (refer chapter 15). It enters into new body already developed in the mother's womb at 7th months of pregnancy. This makes the body for independent survival after delivery. Before 7 months body is unable to survive independent off the body of the mother.

CHAPTER SEVENTEEN

SHRADDHATRAYA VIBHAG YOG
FAITH IN GOODNESS, PASSION, AND IGNORANCE
(SATVAH) (RAJAH) (TAMAH)
SCIENCE OF BREATH

Abstract: The name of this chapter suggests that it deals with division of faith (SHRADDHA-श्रध्दा). According to Gita the GUN or modes of nature are inborn. As per the inborn GUN the nature of person differs. Dominant natural tendency decides nature and behaviour of a person. In previous chapter it is clarified that one who fails to follow the scriptures' laid down in VED, belongs to ASURA or demonic nature. Those who follow belong to DEV or demigod or pious nature. Faith needs acceptance. Acceptance depends upon he temperament of individual. But virtues or GUNAS are inborn but subject to change by cultivations-good or bad. If GOD is imagined for worship and placing one's faith in goodness, passion, or ignorance, do such persons achieve perfect knowledge and reach stage of perfection of life? To satisfy the query of Arjun Shri Krishn explains that there are three types of faith described (in Gita by Shri Krishn).

Trividha bhavati shradha dehinam sa swabhavja I
satviki rajasi cheiva tamasi cheti tam shrunu II 2 II

(Gita. 17. 2).

To understand the stanza let us analyse first. Food body (Annamay sharir-अन्नमय शरीर) depends upon food. Different types of bodies have been described earlier. Type of food consumed influences the mind and mind-set of the person. This was well known to vedik civilization. In fact, the mode of nature to which person belongs decides the outcome. Accordingly, the faith can be grouped into **goodness, passion, and ignorance (satvik, rajasic, tamasic).**

Yajante satvikadevan yaksharakshansi rajasah I
pretan bhutagananshanye yajante tamasa janah II 4 II

(Gita 17. 4).

Those who develop faith in Goodness worship gods and demigods, who have faith in passion worship demons while who have faith in ignorance worship ghosts and spirits:

Quality of pious and demonic natured persons have been described in previous chapter. Persons with three different modes as above, **consume diet which suits them.** Modern science has been advocating different diet items in disease and health. A lot of importance is laid by dieticians for reducing blood Cholesterol and blood sugar for Diabetes and Heart disease (Heart attacks). Age old medical system established role of diet in management of diseases by CHARAK in CHARAK SAMHITA. Interestingly, diet has been described in GITA. The spirituality too depends upon the diet consumed. The role of dieticians in attaining three different modes of behaviour:

Aaharstvapi sarvasya trividho bhavati priyah I
yadnyatapahstatha danam Tesha bhedmimam shrunu II 7 II

(Gita 17, 7).

1. Those who have faith in **goodness**, like healthy diet, which is nourishing and can provide happiness, strength and health. Food of such diet is juicy, fatty, wholesome ad heart pleasing. (GITA 17-8).
2. Those who have faith in **passion**, prefer food, which is bitter, sour, pungent, salty, hot and dry. Such food cause misery, distress and disease. (GITA 17-9)
3. While those who are **Ignorant**, individuals, enjoy food left by others, which is stale, tasteless, putrid and decomposed. (GITA 17, 10).

Indication of this pattern of food habit decides the mentality of a person as well as the mental development and state of mind. Modern science thinks only the physical health, growth and development as far as the food is concerned. What effect the food has on mind is still a topic of research.

Aayuhsatvabalarogyasukhapritivivardhanah I
rastyah snigdhah sthira hrudya aaharah satvikpriyah II 8 II
(GITA 17. 8).

Food is essential for existence of life, for duration of life, strength in life, health in life, happiness in life, satisfaction in life, endurance in life, pleasing to heart in life. This is clearly a Vegetarian food.

Vegetarian food is recommended so that the killing of innocent animals is avoided and nonviolence can be practiced. Animal fat (SNIGDHA PADARTH) is available from milk, Cheese, Butter, Paneer etc. Killing of animals indicates brute mind set.

Devdivajagurupradnyapujanam shouchmarjavam I
brhmacharyamhinsa cha shariram tap uchchyateII 14 II
(GITA 17, 14).

(brahmacharyam, ahinsa = Celibacy, Nonviolence).

It is desired from a person that he should respect and in fact worship elders, Mother and Father, Teachers, Pious knowledgeable individuals. Must follow cleanliness both, externally and internally. He should practice celibacy, simplicity. In fact these are penances of the body of high category (or Sharirik tap – body penance). At the same time, the penance of mind and mind-set must also be practiced:

Manah prasadah soumyatvam mounamatmavinigrahah I
bhavsanshuddhirityetattapo manas uchchate II 16 II

(Gita 17, 16).

Mind-set must be for control of behaviour, purity, satisfaction, simplicity, etc. are said to be qualities for the austerity of mind.

After explaining the importance of diet Lord Krishna explains scientific facts about soul is basically self or ATMA.

Om tatsaditi nirdesho brahmanstrividhah smutah I
brahmanasten vedashcha yadnyasha vihita purah II 23 II

(Gita 17, 23).

"Om Tat Sat" (Supreme that is eternal Truth) are the three words indicated or used to indicate the supreme soul in Sanskrit literature. This is from the start of creation of universe. These words are used in Ved, Yadnya etc.

For this the stanzas of CHHANDOGYOPNISHAD in Chapter 6, Khand 2, shlok 1 and Khand 8, shlok 7 are useful:

Sadev soumyedmagra aasidekmevadvitiyam I
taddheik aahursdevedamagra aasidekmeva dvitiyam
tasmadasatah sajjayat II 1 II

In the beginning the BRAHM was the only one TRUTH (sat\) existing. In true sense, it was unmanifest untruth (asat\) i.e. formless. From this formless the manifest form originated or generated (Tasmadsatah sajjayat).

Sa ya esho nimeitdatmyamidam sarvam tatasatya sa aatma tatvamasi shvetaketo iti bhuya eva ma bhagvanvidnyapayatviti tatha soumyeti hovacha II 7 II

(Chhandogya Upanishad 6, 8, 7).

The micro energy is Aatma. That is the truth. This is SELF. The whole universe is enlivened by this soul. That truth is in every living being. (This statement is by Aaruni to Shvetketu – hence Aaruni says to Shvetketu that "The same is you" –(aatma tatvamasi). In Prashnopnishad 4, stanza 10, it is very nicely explained:

Paramevaksharam pratipadyate say o va ha veitadachchhayam shariram lohitam shubhramksharam vedyate yastu Soumya I sa sarvadnyah sarvo bhavati tadesh shlokah II 10 II

(Prashna 4, 10).

One who knows Shadowless, bodyless, colourless, endless PRAN, ultimately joins Supreme soul, becomes knower of everything or in other words he becomes supreme soul himself, such is the order of this stanza (Shlok).

Gita further explains that the **charity performed** by an individual can be of three modes which are mentioned above – goodness, passion and ignorance. Charity, Penance, Sacrifice, Food intake all advised to be performed in mode of goodness. If performed in mode of passion or ignorance are inferior in quality.

Katvaamla lavana tyushna rukshavidahinah I aahara rajsasyeshta dukhashokamayaprada II 9 II

(Gita 17. 9).

Bitter, excessive citrous, salty, very hot, dry, food generates acidity, diseases, and grief.

Yatayamam gatarasam puti paryushitam cha yat I uchchhishtamapi chamedhyam bhojanam tamasapriyam II 10 II

(Gita. 17. 10).

Stale, tasteless, used by others, food is loved by ignorant persons.

Similarly penance and YADNYA are also described. A stress has been laid on body austerity. According to Patanjal yog Darshan, "Sadhan pad", stanza 1.

Tapahswadhyayeshvarapranidhanani kriyayogah II 1 II
(Patanjal yog darshan Sadhanpad).

Tolerance towards adversities in life, Self-study and devotion to super soul is part of KRIYA YOG. Food, action, gain of knowledge all be dedicated to super soul. This has been preached in Gita Chapter 9. Stanza 27.

Yatkaroshi yadashnasi yajjuhoshi dadasi yat I
yattapasyasi kounteyatkurushva madarpanam II 27 II
(Gita 9. 27).

Whatever one does, eat, or give away, and the auspicious or austerities performed be dedicated to super soul. At the same time certain things are not to be abandoned mentioned in Gita chapter 18.

Yadnyadantapahkarma na tyajam karyamev tat I
yadnyo danam tapashcheiva pavnani manishinam II 5 II
(Gita 18, 5).

The Kriya yog is science of breath. This has been explained previously. Austerity of body helps in evolution of consciousness. Using breathing techniques pranik energy is stimulated to achieve consciousness by controlling life force – PRAN. Interested reader may visit Patanjal Yog Darshan for breathing techniques. Along with control of Pran, other things are also important. Place where to sit is selected properly. (sitting site is Aasan-) The position of spine, neck head be maintained in a straight line. The focus of mind be at the tip of nose. It is also important. In Gita Chapter 6, stanza 11 – 13 it has been described:

Samam kay shirogrivam dharayannachalam sthirah I
samprakshya nasikagram swam dishchanavalokayan II 13 II
(Gita 6, 13).

Prashantatma vigatbhirbrahmacharivrate sthitah I
manah sanyyamya machchito yukta aasit matparah II 14 II
(Gita 6, 14).

This means if one controls mind set (highlighted above), with Aasan and pranayama, with ascetic attitude, totally devoted to super soul in life, the possibility of achieving super consciousness becomes easy.

All this was well known to common public of Sanskrit speaking civilization. Therefore, the example is quoted here by Shri Krishna to emphasize the answer raised by Arjun.

In summary, sacrifice, charity, penance any of such thing performed without faith (SHRADDHA) in super existence, is useless (ASAT). It does not help here in this life nor in life after death:

Ashradhya hutam dattam tapastataptam krutam cha yat I
asadityuchyate parth na cha tatpretya no ihah II 28 II
(Gita 17, 28).

CHAPTER EIGHTEEN

MOKSH SANYAS YOG
RENUNCIATION (SANYAS), ABONDONMENT
(TYAAG)
BIOLOGY OF BEHAVIOUR AND WORK
BIOILLUMINATION

Abstract: The current chapter is basically cannot be considered as a summery as it is thought. In this chapter it mentions some great facts of Zoology and a great philosophy about work and ACTION. Arjun wanted clarification of doubt about Renunciation (Sanyas) and abandonment (Tyag). Abandonment of material desire and activities performed to achieve goal and sense satisfaction (kamyanam krmana nyas) is considered by sages and is termed as renunciation or SANYAS. Giving up results of one's work (sarva karma phala tyag), is considered as abandonment or TYAG. The renunciation, Penance and Charity, Lord Krishn advises caution that these actions are not to be abandoned and described as of three types as per stanza 3, 4 and 5, of 18th chapter, the three modes are meant for purification of great souls.

Any work or "Yadnya karm" (Yadnya, daan, tapah, karm) meant for the betterment of human societies is not to be abandoned. With such actions even the illumined souls also become pure:

Yadnya daan tapah karma na tyajyam karyamev tat I
ydnyo daanam tapahcheiva pavanani manishinam II 5 II

(Gita 18, 5).

A person engaged in duties assigned and those duties performed skill fully with determination, and enthusiasm, along with faith, belongs to SATVIK persons. (or goodness). He has no doubts about the work. Embodied person in material world cannot be without work even for a second. Work cannot be abandoned. According to VEDANTA, there are five causes for the accomplishment of all actions.

Adhishthanam tatha karta Karanam cha pruthagvidham I
vividhashchya pruthak cheshta deivam cheivatra panchamam III4 II

(GITA 18, 14).

Place (adhishthanam), Doer-performer (karta), Instrument (karanam), different endeavours (pruthagvidham), and Supreme soul (deivam) are the five factors of action. Whatever the work performed by the embodied person, by mind or speech which may be good or bad, right or wrong is the result of these five factors. This is a scientific Iand analytical aspect of work and actions. In fact, Aatma is doer and knower both. This has been explained in Prashnopnishad 4, 9:

Esha hi drashta sprashta shrota ghrata rasayita manta
bodha karta vidnyanatma purushah sa perekshare aatmani
sanpratishthite II 9 II

(Prashnopnishad question 4, Shlok 9).

He is watcher, toucher, listener, smeller, taster, thinker, knower, doer, scientific soul. Supreme soul dwell in eternal soul.

Sarvabhuteshu yenekam bhavmavyamikshate I
avibhaktam bhikteshu tajdnyanam viddhi satvikam II 20 II

(GITA. 18. 20).

Spiritual nature is in every living entity. It is one and is in all. Even then they are divided into many forms. What are these different innumerable "forms".

To understand the hidden meaning of this statement, one must refer to concerned literature. In fact, **14 different types of Living entities** are mentioned in Sanskrit literature.

Chaturdasha vidho bhutsargah I

(18ᵗʰ stanza of Patanjal Yog Shastra. (Details are in Patanjal Yog Pradip page 97).

The 14 types of forms are grouped into main three forms – defined in following stanza:

Ashtavikalpo deivsteiryagyonishcha panchdhabhavati I
manushashyeikvidhah samasato bhoutikah sarga II
urdhvam satvvishalstamovishalashcha mulatah sarga I
mdhye rajovishalo brahmadistmba paryantah II

Living beings are grouped or composed in different SARG (रचना).

1. 8 are (ashtavikalpo) **DAIV SARG,** these include: Brahma, Prajapattya, AINDRA, DAIV, Gandharv, Pitrya, Videh, and Prakritlaya.
2. 9ᵗʰ is **MANUSH SARG** (MANUSHI SRISHTI),
3. rest five are **TIRYAK SARG** (animals, birds, reptiles, insects and plants etc.)

BHUH – Living beings on earth, BHUVAH: stars and constellations is Antariksha lok (tara lok), SWAH – Mahendra lok or swarg lok, MAH – Prajapatya swarg lok (Creator), JANAH – TAPAH–, SATYAM – are names of seven (bhuvan) or dwelling places above earth. Last three i.e. Janah, Tapah, Satyam, are three BRAHMA LOK. (see stanza below). Svah, mah, janah, tapah and satyam are 5 swarg or dyoulokah

According to VYAS BHASHYA special places also known as "BHUVAN" where the living being dwell. Earth etc are seven "lok". Hell or PATAL are seven. PRITHVI or earth is BHU LOK. Between earth and POLE STAR (dhruv) where Planets, nakshtra, stars are present, is BHUVAH LOK. Beyond this is SWARG LOK (five types –) or also known as SWAH, Rest 4 LOK are MAH), JANAH lok, TAPAH and SATYAM, Last three (brahmyastribhumiko lok) of these swarg are called as BRAHM LOK. Description of these is in following stanza (in reverse order starting from BRAHM lok to earth):

Brahmyastribhumiko lokah prajapatyastato mahan I
mahendrashcha swarityukto divi tara bhuvi praja II

Vyas Bhasya on sutra 26 of Vibhutipad, adopted from patanjal yog Pradeep pages 523 – 527.

Interestingly the modern science has recently found **"super earth"** (Science 30 May 2017) about 21 light years away and could support life. Probably the search is in BHUVAH lok. The research paper has been published in Astronomical Journal, author Dr. A. H. Martin. University of Canterbury (UC).

Not only living beings but their dwelling places are also described. Modern science of zoology recognizes only Viruses, Micro-bacteria, insects, animals and man. Living entities below the 5th living entity i.e. (5th MANUSH SARG) man, are evident but the eight higher than man are non-evident because of micro nature. Analytical details are in BRIHDARANYAK UPANISHAD, SHATPATH BRAHMAN, TEITARIYA UPNISHAD. The TEITTARIYA upnishad has described in BRAHMANAND VALLI Section, about manifestation of Space (Aakash), Wind (Vayu), Fire (Tej), Water (Drav), and earth (Prithvi) etc. The stanza is like this:

Om brhmachidapnoti param I tadeshabhyukta I
satyam dnyanmanantam brhma I
y oved nihitam parame vyoman I soshnute sarvankamansaha I

brahmana vipashchiteti II 1 II
(Teittariya Upnishad, Brahmanand valli Anuvak 1, stanza 1-2)

Manifestation of Aakash from brahma (or atma), from Aakash Vayu, From Vayu fire or agni, from fire water and from water the earth has come into being. This has already been discussed in prajotpatti, previously. Brahma is omniscient and omnipresent. In our body empty space is in every tissue and organ. It is in heart too. Brahma is in space of heart. Space has come in existence from Brahma. This fact has been researched out by ancient sages of Sanskrit civilization. As per modern science the space is existing since beginning but modern science accepts manifestation of wind from space. Modern science considers the manifestation of fire from wind. Initially wind was cold but due to pressure and gravitational force the heat was generated. Due to heat and other factors the planets and stars got manifested. With cooling effect the water was manifested. After which the earth came into existence. This is what the stanza of upnishad speaks. Next stanza talks about the manifestation of living beings:

Sa va esh purushonnarasamayah I tasyeidameva shirah I
ayandakshinah pakshah I ayamatma I
idam puchchham pratishtha I tadapyeshah shloko bhavati II3 II
(Teittariya Upnishad, Brahmanand valli Anuvak 1, stanza 1-2)

Full of juices the food manifested from earth. All living beings were manifested from food (purushonnarasamayah). Here PURUSH means living body consisting of 75% water and 25 % organic matter which is food. This is what modern science considers. Sanskrit literature further explains that the living one has Head, a right side, a left side, a main part. The main is "I". This "I" is aatma, and a hind end. All universal beings have these body parts. This has been explained in the stanza.

Little meditation on this statement the reader will realize the scientific fact in the stanza. It is further said that the living beings are from food, live and grow from food and become food ultimately is the statement of Sanskrit sages.

Coming back to 14 types of living beings, though it is very categorically stated in Sanskrit literature, at present these are beyond modern science. Because DAIV SARG – living entities are non-evident due to their micro nature not visible to eyes (अदृष्य स्थिति). Science accepts micro-bacteria and viruses because these are evident under microscope. Similarly, the micro living entities described might have been evident to that civilization using some method which we are unaware. But certainly, this is the subject of research, as the statement is very categorical. **The world and creation of 14 types of living beings is related to different KARMPHAL – i.e. result of deeds or results of actions performed in previous incarnation. It is evident from this that the theory of reincarnation is strongly believed and advocated in Sanskrit literature.**

But a person who has engaged himself in renouncement of fruits of work, is truly renounced and totally surrendered to supreme existence, is bound to get liberated from birth-death-birth cycle. Living beings are part of energy from super soul and exist as manifestation of Living embodied life in this material world. Living entities can be grouped into two, one eternally conditioned and the other eternally liberated. The manifested material world repeatedly appear/manifest and repeatedly disappear/unmanifest. Time factor is eternal and governing this celestial drama. All is dependent upon the supreme soul.

Ishwarah sarvabhutanam hruddesherjuna tishthati I
bhramyan sarvabhutani mayaya II 61 II

(GITA 18, 61).

Fraction of super soul energy is situated in the heart of all living beings in the form of soul. This energy directs the material body of living beings. Material world is camouflaged and as a result reality is hidden. Under the effect of camouflaging, apparent reality is assumed as truth by living agencies. Only a few can understand the truth in real sense. The myth, delusion or phantasm causes living agencies to accept the material world as reality. The truth is only revealed to surrendered or renounced seeker.

He only can get transcendental peace and eternal abode and happiness. This explanation given to ARJUN by Lord Krishna and also informed the freedom which all living beings enjoy. Every person works with the genetic predisposition of his mind-set without any outside instigation. Due to affection towards relatives and friends, kith and kin ARJUN is refusing to fight (Gita 18. 60).

Swabhavjen kounteya nibddhah sven karmana I
kartum nechchasi yanmohatkrishyasyavashopi tat II60II
(Gita 18. 60).

In other word, Arjun is bound down due to genetic predisposition to fight the war forced upon him as his own DHARM – the basic instinct of being a KSHATRIYA – fighter. At the same time despite the freedom available to all living beings, Lord Krishna suggests that deliberately ponder and understand only then do the work if one wishes to do. In the same context father of ATOM BOMB scientist Oppenheimer did his job as a scientist.

How simple the example sounds but scientifically the whole genetics has been put forth. The hidden description of science of genetics has been narrated to the reader. This is because of the common scientific knowledge of the SANSKRIT speaking civilization was very high. Shri Krishn further endorses that Arjun would definitely fight with his **genetic instinct.** (GITA 18, 63). Described the secret knowledge, but he has to decide if he feels to fight. It is not because of the instigation to fight. KRISHNA has narrated the secret knowledge, but ARJUN'S own decision would be final.

Iti te dnyanmakhyatam guhyad guhyataram maya I
vimrushyeitdasheshen yathechchhasi tatha kuru
(Gita. 18. 63).

The confidential knowledge cannot come to "mind set" unless one is pure, self-controlled, unattached devoted or engaged in and surrendered totally to devotional service.

Whatever is coming to mind because of basic state of mind of a person, depending upon the development of, and genetic development of DHARM (not religion) of an individual, KRISHNA advises to abandon the other DHARM (not religion) coming to mind, and follow Shri KRISHNA. This is, he explains that if one follows other DHARM – instinct, it would be a death like outcome for the person. Therefore, follow Supreme soul as KRISHNA and HE (paramatma) only, can help to overcome sinful effects of the work involved. In other word, Total surrender to supreme soul, without fear, can free a person from effects and results of sinful acts.

Sarvadharman parityajya mamekam sharanamvraja I
aham tvam sarva papebhyo mokshayishyami ma shuchah II 66 II

(GITA 18, 66).

As a warrior when a war is forced upon ARJUN, killing of rival forces is essential act which is evil act, but duty performed is lawful (laws of nature). Such acts, if dedicated to super soul without any desire or hope for the result outcome, are free from bondage. The evil act will not affect the doer.

(A nice example of modern day event is "Surgical strike" on neighbouring country territory due to killings of innocent population by terrorist, is definitely free from evil results of action)

Despite the control by super soul and genetics provided (i.e. beyond individual control), to living beings still the actions and deeds are under individual living being. This way the deeds and actions are under individual control. To set the "mid set", pious or evil, is prerogative of individual. Hence SANJAY an eye-witness, of whole discourse/ GITOPDESH, said at conclusion that:

Yatra yogeshwarah krushno yatra partho dhanurdharah I
tatra shrirvijayo bhutirdhruva nitirmatrmam II 78 II

(Gita 18, 78).

YOGIK mind-set (of KRISHN) and action executor (like ARJUN), the success is inevitable.

In conclusion, the creator manifested creation and creature to satisfy HIS (supreme-soul), ulterior motive, unknown to us, in the manifest universe. Harbouring the false ego, we consider ourselves, as charioteer of our lives. In fact, the charioteer of the creation and creature of manifest universe, controlling everything and the manifest universe in incessant stream till eternity is unknown to us. But can be imagined in mind. We the humans, with false ego and dreamy state of our minds, fall prey and succumb in the falsely and camouflaged manifest universe, seemingly real in every aspect. All is the game of the mind. We are tools of mind which controls and compels manifest bodies which are tuned to play characters of drama of creation till eternity.

ABOUT THE AUTHOR

DR Balkrishna Matapurkar

Pioneering research First time in the world. Regenerative Surgery. Regeneration of tissues and Organs in the body from Adult tissue Stem Cells is for the first time in the world, **a gift from India to the world,** is a pioneering research by Dr. B.G. Matapurkar. It is a very cost effective technique. No donor, no tissue matching, no postoperative costly drugs to suppress immunity. This is in continuation with the tradition of culture of selfless Rishi Muni's of "Bharatvarsha" (India)

Invented a new Physiological phenomenon of Mammalian body by which human body can regenerate its own tissues and organs. Named as Desired Metaplasia, now recognized universally, and techniques published in Text Books on medical subjects.

International Patents. This research got International patent from USA in 2001 (effective from 1995), with the help of TIFAC, branch of Department of Sc. and Technology, Govt. of India, in the name of M.A. Medical College, New Delhi. **The patent was awarded to Dr. Matapurkar by the then Hon'ble Science technology and**

Education minister Shri M M Joshi in a Press Conference organized by Govt. of India. (U S Patent and Trademark Office. United States Patent No. 6227202 dated 2001 and 20020007223 dated 2003).

Unique research for the Humans and mammals (animals). His research has stimulated researchers universally. It is proving beneficial to suffering humanity and mammals. Stem cells are being used universally, in the management of human diseases.

History of Medicine. In the History of Stem Cell Research starting from 1908 till 2013, Stem Cell Engineering Key research Wikipedia Octber 2013, Dr. Matapurkars name has entered in the year 1995 and 1997, as pioneer research on Adult Stem Cell. See Web site Stem cell research by Makayla L, 2014, Prezi.com and history of Medicine-stem cells.

History of Medicine (Human beings) in sections of Modern Surgery. Time line.

Text books on medical subjects. Dr. Matapurkars New operative techniques have been published in Medical Text Books:-

1. R. Maingots Abdominal **Operations** for MS, M Ch curriculum 1997, U S A.
2. Text Book of **Obstetrics and Gynecology, J P Publishers, India 2010.**
3. **New physiological phenomenon of Human and Mammalian body – Desired Metaplasia – Published in Text Books of Physiology and Pathology.**
4. Local Cells, Global Science: The Rise of Embryonic Stem Cell Research in India Genetics and Society. Aditya Bharadwaj and Peter Glasner, 2008
5. **Surgical Research. Wiley W. Souba, Douglas Wayne Wilmore, 2001**
6. **Female Urology: Text with DVD. Shlomo Raz, Larissa V. Rodriguez, 2008**

Honble P M Shri Narendra Modiji has recommended his pioneering research for school curriculum **in Gujrat educational books for Primary and Secondary education in a book** by Historian Dr. Dinanath Batra. Ref. Indian Express news Paper front page 27-10-2014.

ORGANIZER NEWS. Dr. B.G. Matapurkar's research work is like introducing zero to the world of Mathematics by RISHI MUNIs of ancient Bharat vide analytical article Bharat Pioneer of Discoveries by Mr. S Raje in ORGANISER on 14 JUNE 2015.

Special performance: Surgical feat.

1. During his posting in Andaman Nicobar Islands 1973 to 1979, He is successful in using natural resources of the land The **intra venous use of** tender **Coconut Water, after Scientific Research. It proved a life saving measure in needy patients,** in absence/ short supply of Intra Venous Glucose and Saline in remote areas of Islands during 1973-1979. Later on this was Published in Science Journal. Journal of institute of Medicine 1999:21:1-57
    During this period, being the only surgeon available. (During 1973 to 1979 in A &N Islands). He performed difficult life-saving surgical operations successfully when no anaesthetist was posted in A & N islands, like:-

    a. Life saving Difficult cases of Head Injuries,
    b. Difficult life saving Gynecological & Caesarean sections in remote places of A & N islands, as no Gynecologist or Obstetrician was available
    c. Orthopedic surgeries etc. in absence of Orthopedic Surgeon
    d. Spinal and general Anaesthesia before surgery as no anesthetist was available in fact no post was created at Car Nicobar, CIVIL Hospital.

2. **He operated in an emergency patient when he himself suffering from 104 degree temperature. Certificate attached (-photocopy).**

ACHIEVEMENTS

Dr. Matapurkar opted to serve in remotest areas of India Andaman and Nicobar islands (1972) where no surgeon wanted to go and serve. After M.S. degree in surgery and leaving a regular job in Delhi, he desired to serve in A and N islands. He discovered Intra Venous use of tender coconut water (see page 367 below).

Discovered and Established **new PHYSIOLOGICAL phenomenon of human body "Desired Metaplasia" by** which body can regenerate own tissues and organs. Universally accepted and included in Text Books

New regenerative technique evolved using adult tissue stem cells. Patents on the subjects. Patented and now in Text books.

The technology of regenerating tissues and organs in the body, is a cost effective technology, is good for Rich and poor sections of society. It is cost effective because no donor needed, no tissue matching, no costly drugs to prevent rejection etc.

Developed I.V. use of coconut water in human patients after research for use in A and N Islands, due to unavailability of Intra Venous fluids. Many patients of Nicobar Island were benefited and survived with severe dehydration due to diarrhea and dysentery. Later on published in Science journals

NEW OPERATIVE TECHNIQUES

Developed which are included in Text Books of Surgery, Text book of Obst. and Gynecology. Text books of basic medical subjects – Physiology, Pathology etc.

Text Books of Surgery by R. Maingots Abdominal **Operations** for MS, M Ch curriculum 1997, U S A.

Text book of **Obstetrics and Gynecology, J P Publishers, India 2010.**

In fact he has established subject of Regenerative Surgery for the first time in medical history.

FUTURE USE. The Stem cell technology is useful in growing Meat in the lab for human consumption and **for future space travel – growing meat in space ship** low weight cargo and fresh food availability

Special performance Surgical feat.
During his posting in Andaman Nicobar Islands 1973 to 1979, He was successful in using natural resources of the land The **intra venous use of** tender **Coconut Water, after Scientific Research. It proved a life saving measure in needy patients,** in absence or short supply of Intra Venous Glucose and Saline in remote areas of Islands where rarely the ships available for supply of Medicine during 1973-1979. Later on this was published in Science Journal. Journal of institute of Medicine year 1999 Volume 21 pages 1 to 57

During 1973–1979 in Andaman & Nicobar Islands, he performed difficult life saving operations when no anesthetist was available:

- Lifesaving Difficult operations of Head Injuries,
- Difficult lifesaving Gynecological and Caesarean sections operations in remote places of A and N islands, as no Gynecologist or Obstetrician was available
- Orthopedic surgeries etc. in absence of Orthopedic Surgeon
- Spinal and general Anesthesia before surgery as no anesthetist was available in fact no post was created.

He operated in an emergency patient when he himself was suffering from 104 degree temperature. Certificate attached-photocopy by commanding officer Navy.

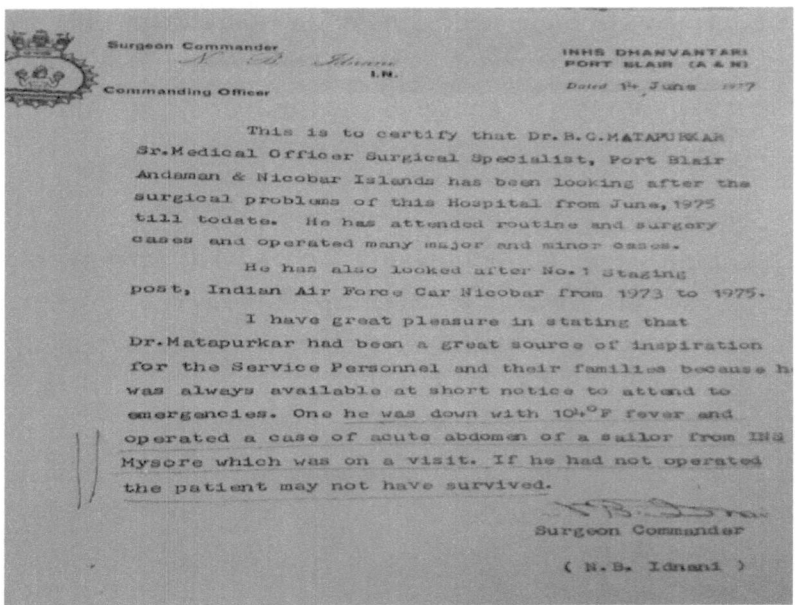

National and State Award. Dr. Matapurkar was awarded by Hon'ble Shri Vinay Sahastrabudhe, Loksabha Member, Shri Jajuji, Vice president B J P, and Ramdas Aathavleji, Cabinet Minister

First ever GRMed College Ratna गजराराजे मेडिकल कॉलेज रत्न **award Presented by Dean, GR Medical college Gwalior, M.P. 7th March, 2019 Photo and press cuttings attached**

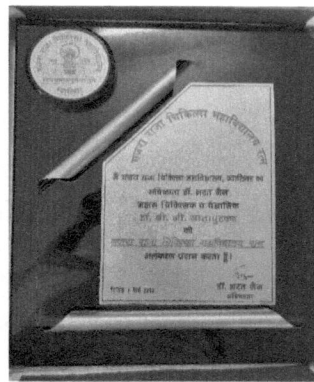

जहां मेडिसिन की एबीसीडी सीखी, वहीं व्याख्यान देना गर्व की बात

विश्व प्रसिद्ध वैज्ञानिक डॉ. मातापुरकर गजराराजा चिकित्सा रत्न से आज होंगे सम्मानित

ग्वालियर, न.सं.। योगिराज हरकरे पुण्य तिथि पर 7 मार्च को गजराराजा चिकित्सा महाविद्यालय, ग्वालियर में विश्व प्रसिद्ध वैज्ञानिक डॉ. बी.जी. मातापुरकर को गजराराजा चिकित्सा महाविद्यालय रत्न सम्मान से सम्मानित किया जाएगा। यह कार्यक्रम चिकित्सा महाविद्यालय में दोपहर 1. 30 बजे आयोजित होगा, जिसमें डॉ. आर.डी. वर्मा मुख्य अतिथि, डॉ. एन.डी. वैश्य व सच्चिदानंद महाराज विशेष अतिथि होंगे। महाविद्यालय के अधिष्ठाता डॉ. भरत जैन अध्यक्षता करेंगे। इस अवसर पर डॉ. बी.जी. मातापुरकर ऑर्गन री-जनरेशन एवं विज्ञान व भगवतगीता पर उद्बोधन देंगे। साथ ही डॉ. मातापुरकर शोधार्थियों को मार्गदर्शन देंगे। यह जानकारी कार्यक्रम के संयोजक डॉ. ईश्वरचंद्र करकरे व संचालक डॉ. भदौरिया ने दी।

Gwalior Edition
7 Mar, 2019 Page No. 3

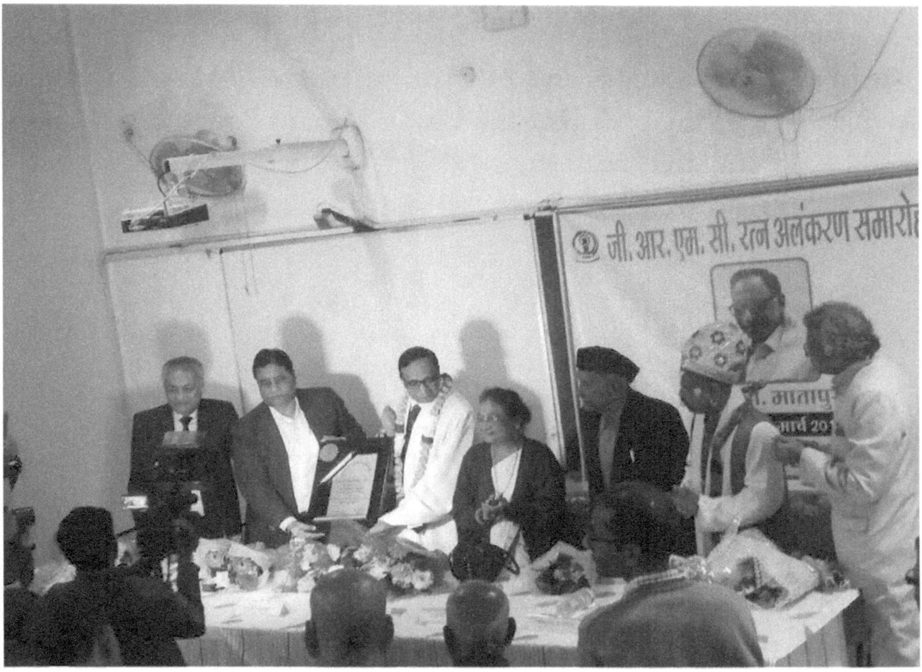

Must watch Video to understand the research. This was aired by National TV channel. TURNING POINT also published on you tube.

Dr. B.C. Roy Award. Drs Day Award Dr. B.C. Roy 2001 by the then Honble Chief Min. Delhi, NCR Shrimati Shiela Dixit **Photo attached**

International Patent award 2001 by the then Honble Cabinet Min. for **Sc Technology & Education, Dr. M. M. Joshi. Photo and press cuttings attached**

SASTRA University TamilNadu. International Oration Award 2002 by Deptt of Biotechnology, SASTRA University, Tanjavur, Tamilnadu

Gwalior Medical Association Delhi Felicitation Award 2001 by the then **Hon'ble Health Min** of **Delhi/NCR, Dr. Ashok walia.**

Brihanmaharashtrian Utsav Samiti. award 2002 By the then Honble Loksabha Speaker Shri Manohar Joshi at Jaipur, Rajasthan.

World environment Day and 21st Anniversary of N M N H Oration Award 2002 by the then Hon'ble Union Min. of Power, Shri Suresh Prabhu. Photo attached

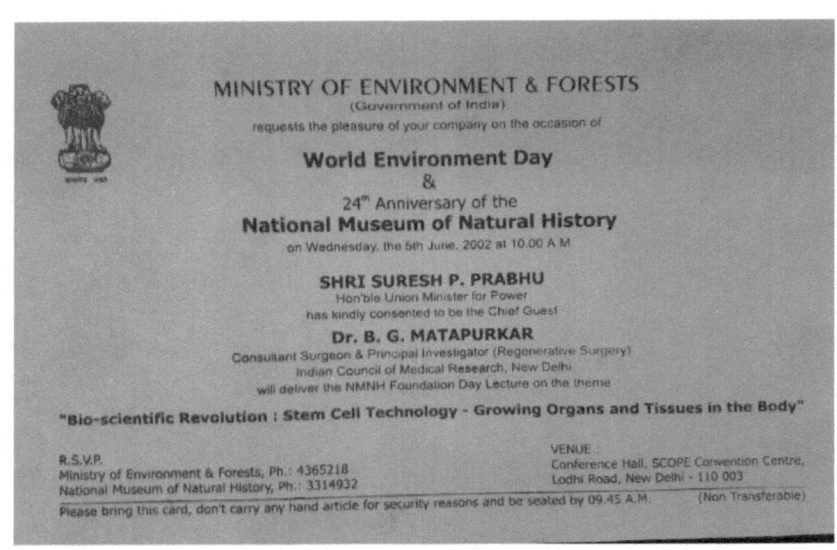

Madhya Pradesh Gwalior Vikas Samiti Award on 21st Abhinandan samaroh 1999, at Gwalior, Madhya Pradesh.. Awarded by Principal, Vikrant College, Gwalior, M.P. 2017.

Delhi Maharashtrian Educational & Cultural Society, Award 2001 by the then Hon'ble Min. for State Govt of Maharashtra.

Prestigious since 1931, 90 years old Sharad Vyakhyan Mala Oration Award 1997 Gwalior, Madhya Pradesh. Topic VED VIDNYAN and Chikitsashastra. Many eminent laureate persons have been invited to this highly acclaimed lecture series Prof. P.K. Atre, Prof. P.L. Deshpande, V S Khandekar etc.

Madhya Pradesh Gwalior Vikas Samiti Award on 21st Abhinandan samaroh 1999, at Gwalior,

Madhya Pradesh. Awarded by Principal, Vikrant College, Gwalior, M.P. 2017.

डॉ. मातापुरकर का सम्मान

पत्रिका PLUS रिपोर्टर

ग्वालियर ● मानव शरीर के अधिकांश अंगों एवं ऊतकों को पुनर्उत्पादित किया जा सकता है। यह एक नियमित प्रक्रिया है। यह सम्पूर्ण ज्ञान हमारे शास्त्रों शिवपुराण, दुर्गापुराण, महाभारत, भागवत में उपलब्ध है। यह जानकारी डॉ. बीजी मातापुरकर ने पार्टिसिपेंट्स को वर्कशॉप के दौरान दी।

यह वर्कशॉप विक्रांत ग्रुप ऑफ इंस्टीट्यूशंस की ओर से मंगलवार को परिसर में आयोजित की गई थी। इस दौरान डॉ. मातापुरकर का सम्मान भी किया गया। कार्यक्रम में शिक्षाविद् डॉ. पीएस बिसेन, कॉलेज के चेयरमैन आरएस राठौर, सचिव विक्रांत सिंह राठौर, वॉइस प्रेसीडेंट संजीव सिंह चौहान, रजिस्ट्रार रिचा वर्मा सहित सभी फैकल्टी उपस्थित रही।

Award presentation by: 1. ShriManohar Joshiji, Speaker Lok sabha, 2. Shri M M Joshiji, Sc & technology and education minister, 3. Shri Suresh Prabhuji, The then Sc and Technology Minister, 4. Shrimati S. Dixitji Chief Minister Delhi, 5. President, American college of surgeons in AIIMS Delhi 6. Education Minister, Maharashtra State

International Awards

1. **Elected Active Member. New York Academy of Sciences**. Membership Award of New York Academy of Sciences, U S A, 1994 as

2. Fellowship Award of **Marquis Whos who Publication Board 1997, U S A**

3. American College of Surgeons India Chapter Oration award 1999 by **President of A C S, U S A. See photo above**

4. Life time Membership Award of **American Biographical Institute Board of Governers 1998, USA**

5. **Limca Book of World Record**. World Record Published in Limca Book of Records 2002, page 85 Regeneration of Ureter in Live Mammalian Body for the first time in the world

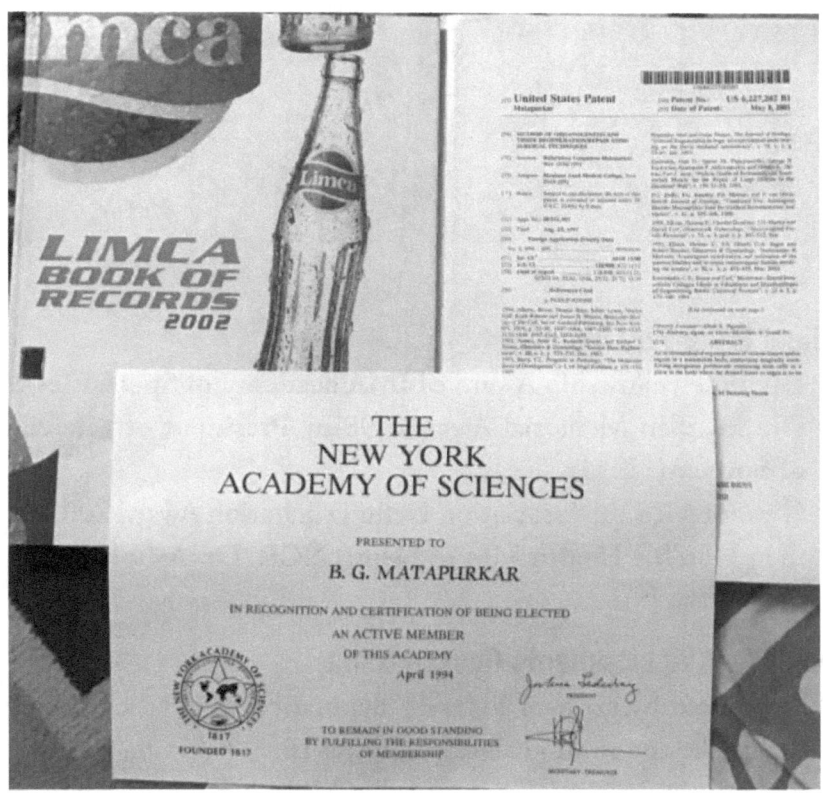

Elected active member of NYAS.

6. **A press release in USA in INDIA ABROAD after a lecture in New York Academy of Sciences at Bathesda, Wasington DC, USA, 1997.**

Paper on organ regeneration presented

By a Staff Writer

NEW YORK — Dr. B. G. Matapurkar, a medical researcher, presented a paper on regeneration of tissues and organs at a conference of the New York Academy of Sciences in Bethesda, Maryland, last month.

The paper was based on what Matapurkar calls his "breakthrough research" on the subject. He is the principal investigator of the "Project Regeneration Surgery" of the Indian Council of Medical Research, and a consultant at the Department of Surgery at the Lok Nayak Jai Prakash Hospital and Maulana Azad Medical College, New Delhi.

Dr. B.G. Matapurkar, who presented his research paper at the conference of the New York Academy of Sciences.

According to the paper, Matapurkar, along with Dr. Harmeet Singh Rehan, a senior resident at the college, have regenerated the ureter, the fallopian tube and the uterus in laboratory monkeys and dogs. The research was conducted on the basis of "Matapurkar's Hypothesis" that peritoneum stem cells can be converted into desired tissues and organs in the body under proper environment. His objective was to find an alternative to organ transplant which has limitations like nonavailability of organs, rejection in the body and immunological complications.

Matapurkar, who is an elected member of the academy, has been working on the regeneration concept for the past 20 years, published report quoted him as saying. He said his work was based on "embryological principle." The regeneration of organs is possible through a simple surgical technique, he said.

Rehan was quoted as claiming that the technique could be performed by an average trained surgeon in a small hospital. The technique also would prove a boon in the management of diseases, he said.

India Abroad., New York ed : Oct. 97 - FRI 14 Nov 97.

Medical Fraternity Awards

1. Life time Fellowship Award **of IMA academy of Specialties 1996**
2. Dr. S.K. Sen Memorial Award1998 by **President of Association of Surgeons India.**
3. Gwalior Medical Association Delhi Felicitation Award 2001 by the then **Hon'ble Health Min** of **Delhi, NCR, Dr. Ashok walia.**

Special Task Assignments Govt Of India

1. Appointed by govt. of India on deputation to govt. of Lebnon as deputy leader of Surgical team during Israel Palestine war, to clear war injuries in 1981 Govt. of India Min. of External affairs office order dateda.ed 25th July 1981.

2. Appointed by Govt of India as leader of Surgical team to USSR, in 1988 at the time of Earth Quake Disaster in Soviet Russia.

3. Posted for 45 days to Dadra Nagar Haveli, Silvasa, to coincide with the visit of President of India Govt. of India Min. of Health and family Welfare, Dept. of Health ND. Office order dated 28th September, 1990

4. Appointed as Surgical Expert & shift In Charge for 37th Commonwealth Conference 19th to 29th Sept. 1991 at Mini Hospital, Ashok Hotel, ND, India.

5. Invited to help setting Museum Section on stem cell technology by NMNH, Govt of India 24th Anniversary dated 5th June, 2002.

6. Publications: Many research papers have been published in Reputed Science Journals after peer review. To name a few:

    British J Urol., Ind J of Cancer, World J of Surgery 1991 and 1999, International Chirurgica Dige. ASAIO journal(USA)

कलयुग के ब्रह्मा | बातें सबके साथ httpwww.rrtd.nic.in ddjune2k1.htm

http en.wikipedia.org wiki B.G._Matapurkar

http www.pharmabiz.com article detnews.asp Arch articleid 7705 sectionid 44

Congress CICD Iternational J of Hospital Medicine, Ind J Experimental Biology, 1996, 2000, 2001, 2002, 2003, 2010, Ind J of Urol, 1996 1997 J. of American College of Surgeons, Morphogenesis –cellular interactions, N Y A S, USA. ASAIO Journal – American Society of Artificial Internal Organs Journal. Polish Medical Iournal. Molecular Biology – American Society for Cell Biology, Matapurkar's Method of Surgery on Abdominal Hernia, Chirurgica Ogolnej. Nanotechnology and Health care J of SASTRA University Tamilnadu.

Modified Sandwich technique in Incisional Hernia: Annals of Colorectal res. 2013: DOI 10.5812/12/acr.11481. Sandwich technique developed by B G Matapurkar in 1991 1nd in 1999

International BIOGRAPHICAL PUBLICATIOS: Various International Biographical publications:

a. **Marquis Whos who in the World** in 14-17th ed. Whos Who in Medicine and Health care.

b. **International Biographical Centre, Cambridge, UK.** International whos who of intellectuals 1997, Dictionary of international Biography, 27th ed. March 1999.

c. **American BiographicalInstitute, Inc.** Invited member of Research Board of advisors.

d. **Reference Asia vol.** In Asias Whos who of men and weman of achievement 1991, Asian American whos who1998,

e. **Rifacimento International** Vol Asia pacific whos who1998

f. **International Penguin** publishing House Rising personalities of India, 1998.

g. **South Asia Overseas publications Co.** Twentieth Century Distinguished whos who, 1997.

h. **History of Pioneers of adult stem cell research Web site Prezi. com to see slides**
   http prezi.com m 9x6chwqfviz Robert hooke

Publications Only few path breaking research. In INTERNATIONAL JOURNALS (Only for reference purpose for interested reader)

1. A critical evaluation of Vasoseminal vesiculography. Original research British J. of Urol. Aug. 1969. Matapurkar, B.G.; Taneja, O.P.; Saha, M.M.; Bhardwaj, O.P. Vol. XLI, p. 455 464.

2. Clinico pathological study of carcinoma prostate, Ind. J. of Cancer. Sept. 1969. Matapurkar, B.G.; Taneja, O.P. page. 172 183.

3. Re-canalization of vas deferans A thirteen years study. Presented in International conference on Voluntary Sterilization & Family Welfare, Organised by govt. of India, 1986. Matapurkar, B.G.

4. New technique of repair of incisional hernia. Presented the paper in a Congress International Chirgica Digestiva CICD, 1990 held at New Delhi. Matapurkar, B.G.

5. New technique of Marlex peritoneal sandwich in the repair of large incisional hernia. World J. of Surgery, 1991, 15, 768-770. Matapurkar, B.G. Gupta, A.K. Agarwal, A.K.

6. Large Abdominal Incisional Hernia, Hospimedica, An international J. on Hospital Medicine, Jan, 1992. Matapurkar, B.G. Vol. 10, no. 1, p. 36 to 41.

7. Formation of neoureter from peritoneum in live animal model dog. Indian J. of Expt. Biology. 1996, 34 954 to 958. Matapurkar, B.G. Rehan, H.M.S.

8. Organogenesis and tissue regeneration of ureter from autogenous primitive stem cells in adult Rhesus monkeys, Indian j. of Urol. 1996, 13 pages 1 to 4. Matapurkar, B.G.; Hermit Singh Rehan, Agarwal, A.K.; Anand Kumar.

9. New technique of repair of Complex genito urinary rectal fistula using peritoneum for urethral reconstruction and as an inter positional tissue, Ind. J. Urol. 1997, 14 36 to 44. Matapurkar, B.G. Rehan Hermit Singh

10. Regeneration of ureter by desired metaplasia of stem cells. Presented in inaugural conference of American College of Surgeons at AIIMS, ND. 1996. Matapurkar, B.G.

11. Organogenesis and tissue regeneration of ureter, fallopian tube and uterus using stem cells of autogenous peritoneum. Presented in conference at USA, Morphogenesis Cellular interactions. Conference held at Bathesda, MD. 1997. Matapurkar, B.G.

12. Organogenesis by desired metaplasia of stem cells, Annals of New York Academy of Sciences, Vol. 857 p. 263 to 267, edited by R. Fleischmajer, R Timpl, Z. Werb. 1998.
    Author Matapurkar, B.G. Bhargave, A.; Dawson, L. Sonal, B.

13. Regeneration of Abdominal wall aponeurosis New dimension in Marlex peritoneal Sandwich repair of incisional hernia. World J. of Surgery, 1999, 23 446 to 451.
    Matapurkar, B.G. Bhargave, A. Dawson, L. Sonal, B.

14. Organogenesis and Tissue regeneration of Fallopian tube A Desired Metaplasia of mesodermal stem cells in live animal models dogs Ind. J. Expt. Biol. 38 129 to 136 year 2000

15. A review article. Basic philosophy of Organogenesis & Histogenesis by Desired Metaplasia of autogenous stem cells in vivo and its Human Utility. WJS.

16. Organogenesis by Desired Metaplasia of autogenous stem cells in vivo. Presented in XXXII World congress of the International College of Surgeons. On 10-10-2000.

17. U.S. International Patent No 6227202 Dt. 8th May, 2001. Method of organogenesis and tissue regeneration Repair using surgical techniques.

18. Neo-organogenesis an Neo-histogenesis by Desired Metaplasia of autogenous tissue Stem Cells In vivo A Critical and Scientific evolution with 125 years of review literature. American Society for Artificial Internal Organs Journal 2002. January issue of 2003.

19. A new Physiological Phenomenon of mammalianbody for Organ and Tissue Neoregeneration in vivo Adult stem cell technology in the perspective of literature. Ind. J of Exptl Biol. Vol. 40 Dec, 2002 pp. 1331 to 1343.

20. Zabieginaprawcze w duzych przepuklinach brzusznych w bliznach pooperacyjnych. Med. Prak. WS 1993 2 38 to 41 POLISH MEDICAL JOURNAL.

21. Matapurkar's Method of peritoneal Sandwich in treatment of postoperative Abdominal hernias Published by Aleksander Sowula. Z. ODDZIALU Chirugii Ogolnej. Szpitala im. L Rydygiera w Katowicach Surgery department of the L. Rydygier Hospital in Katowice Kierownik: Lek. Med. Henryk Groele.

22. Matapurkar, B.G. neo-organo-histo-genesis A Physiological Phenomenon of Desired metaplasia of autogenous Adult Tissue Stem cells. Molecular Biol of the CELL Abstracts. 42nd American Society for Cell Biol. December 14 to 18, 2002 San Francisco, California. N0. 769, pp 137a.

23. Invited Lecture in International Workshop on Nanotechnology and Healthcare, 11th to 12th January 2003 at SASTRA CAMPUS, ARTS, SCIENCE, TECHNOLOGY and RESEARCH ACADEMY, DEEMED UNIVERSITY, THANJAVUR, TAMILNADU, INDIA

Contribution to Society

Opted to serve remotest territory of India 1973 to 1979. During that period no Surgeon desired to serve there. He used local resource of Tender coconut water for Intravenous use for needy patients to save life in remote areas of Islands. The regeneration of Tissues and Organs is a great gift to Society and

Humanity of the world and also to all animal kingdom, Veterinary sciences in the world. The new cost effective technique is useful to all factions of society the Rich and poor. Regeneration is very cost effective as it needs no donor, no rejection, no costly drugs to suppress Immunity as body regenerates tissues and organs using own tissue cells..

Many difficult to treat diseases can be managed with this technique.

Future space travel load can be reduced by growing meat a protein source, in the space laboratory. This research has stimulated researchers universally. It is proving beneficial to suffering humanity.

Delivered Lectures on Gita and Science at:-

PUNE, Maharashtra,

Gwalior, MP,

Tanjavur, TAMILNADU

Hyderabad, AP,

Austin USA,

Bathesda Washington DC USA

see Gita colloquium card below and Sharad Vyakhyan Mala – invitation letter. This is to make aware the society about science in our Vedik literature.

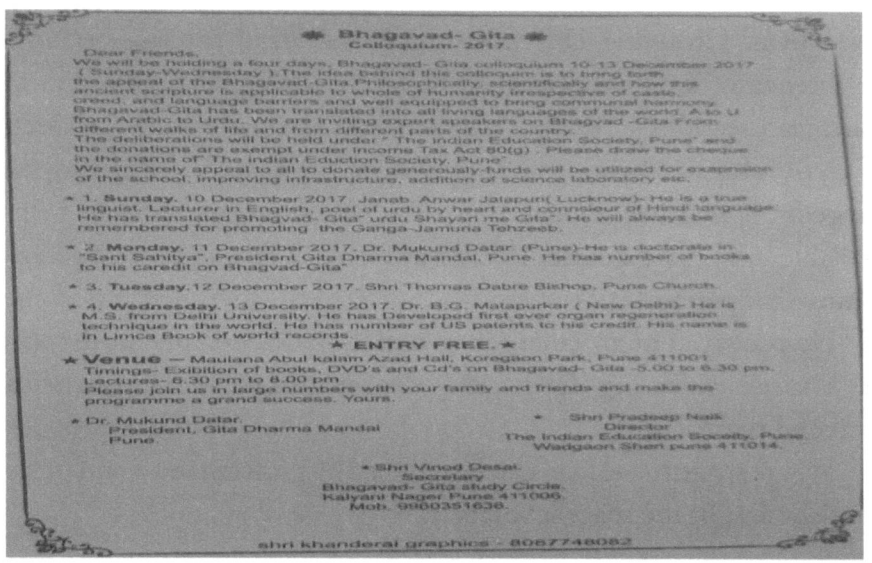

Contribution to field

The technique of regeneration of organs in body is for the first time in the world and has received international patents from USA. **Human use has also been patented from US PTO.** This research has discovered a **New Physiological phenomenon of Human and Mammalian body** Desired Metaplasia for the first time in the world by which body regenerates its own tissues and Organs in the body with the help of body own tissue Stem Cells. It has solved many unanswered questions in medical science It has been recognized and appreciated universally.

The regenerative Surgical Techniques are published in **Text Books of Basic Medical Sciences and Specialized Medical Sciences.** The pioneering research has stimulated researchers universally. It is proving beneficial to suffering humanity. **Cord blood preservation started all over the world.**

Research on coconut water is very useful and Intra Venous use of Tender Coconut water in remote areas of country where I V Glucose or Saline is scanty or not available. In Andaman Nicobar posting being the only surgeon available He performed difficult lifesaving surgical operations successfully when no anesthetist was posted in islands and saved lives. Lifesaving Difficult surgery for Head Injuries performed in absence of **neurosurgeon.** Difficult lifesaving Gynecological and Caesarean sections in remote places of islands, successfully in absence of **Gynecologist** Orthopedic surgeries etc. in **absence of Orthopedic Surgeon.** Spinal and general Anesthesia before surgery with the help of available staff, as **no anesthetist** was available in fact no post was created during that period.

He operated in an emergency patient from defense sector when he himself suffering from 104 degree temperature. Certificate attached photocopy.

Future space travel load can be reduced by growing meat and protein nourishment in the space laboratory.

NEW Sutgical TECHNIQUES developed which are included in Text Books of Surgery, Text book of Obst. and Gynecology. Text books of basic medical subjects Anatomy, Physiology, Pathology, Urology etc.

Text Books of Surgery by R. Maingots Abdominal **Operations** for MS, M Ch curriculum 1997, U S A.

Text book of **Obstetrics and Gynecology, J P Publishers, India 2010.**

Surgical Research Wiley W. Souba, Douglas Wayne Wilmore, 2001. G813 G822. 3. Mason, R. J 1997. Stem cells in lung development, disease, and therapy. Am. J. Respir. Cell. Mol. Biol. 16 vol 4, page 355 to 363. 4. Matapurkar. B. G., et al. 1998. Organogenesis by desired metaplasia of autogenous stem cells.

Female Urology Text with DVD Shlomo Raz, Larissa V. Rodriguez, 2008

Surg Gynecol Obstet. 1988 vol 167 page 124 to 128. Matapurkar BG

OTHER BOOKS ON DR. B.G. MATAPURKAR

Textbook of Gynecology
2011
For web site click on figure

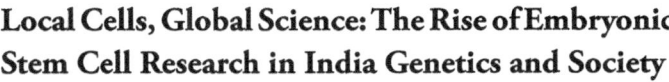

Local Cells, Global Science: The Rise of Embryonic Stem Cell Research in India Genetics and Society

Aditya Bharadwaj and Peter Glasner, 2008

One of the first studies of an exciting new development in global biotechnology, this cutting edge text examines the extent of the transnational movements of tissues, stem cells, and expertise, in the developing governance ramework of India. Documenting the impact of local and global governance frames on the everyday conduct of research.

Local Cells, Global Science: The Rise of Embryonic Stem Cell...

Aditya Bharadwaj, Peter E. Glasner, 2009

Cick on figure for web site

The most extraordinary justification for stem cell research, however, came from a scientist, Dr. B.G. Matapurkar, who conducted research using adult stem cells in Delhi's Maulana Azad Medical College. He argues that Adi Parva first chapter of...

Multiplicity Yours: Cloning, Stem Cell Research And Regenerative Medicine

Hwa A. Lim, 2006

BJP Today

2001

Such claims do not flow from a new age guru, however, but from Dr. B.G. Matapurkar, a pioneer in the field of stem cell research in India. Matapurkur insists that embryonic stem-cell research is one of the lost sciences of India. In the Adi Parva,

Click on figure for web site

The Adventure Time Encyclopaedia Encyclopedia Inhabitants, Lore, Spells, and Ancient Crypt Warnings of the...

Martin Olson and Hunson Abadeeer, 2013

Martin Olson is a comedy writer, television producer, stage director, and composer. Olson is best known as a founding father of the Boston comedy scene, and he has received an Emmy nomination and Ace Award for television writing. Olson lives in Los Angeles.

Click on figure for web site for details

Surgical Research

Wiley W. Souba, Douglas Wayne Wilmore, 2001

G813-G822. 3. Mason, R. J., et at. 1997. Stem cells in lung development, disease, and therapy. Am. J. Respir. Cell. Mol. Biol. 16 4, 355-363. 4. Matapurkar. B. G., et al. 1998. Organogenesis by desired metaplasia of autogenous stem cells.

Click on Figure for web site

Female Urology: Text with DVD

Shlomo Raz, Larissa V. Rodriguez, 2008

Surg Gynecol Obstet. 1988; 167: 124-128. 5O Matapurkar BG, Gupta AK, Agarwal AK. A new technique of MarlexPeritoneal Sandwich in the repair of large incisional hernias. World J Surg. 1991; 15: 768. 51 Molloy RG, Moan KT, Waldron RP,

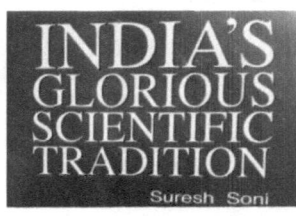

PTI news-Scientific details How 101 children born by Gandhari described by Dr. Matapurkar in an international science conference (2002), published in Chapter 18, Hygiene and Health of the book. Ocean Books Pvt Ltd ISO9001:2008 Publishers.

The following photographs show Awards given to Dr. B.G. Matapurkar BY Hon'ble Shri M M Joshiji, Mr. Manohar Joshiji, Mrs. Shiela Dixitji, Shri Suresh Prabhuji, State Education Min and President of American Association of Surgery

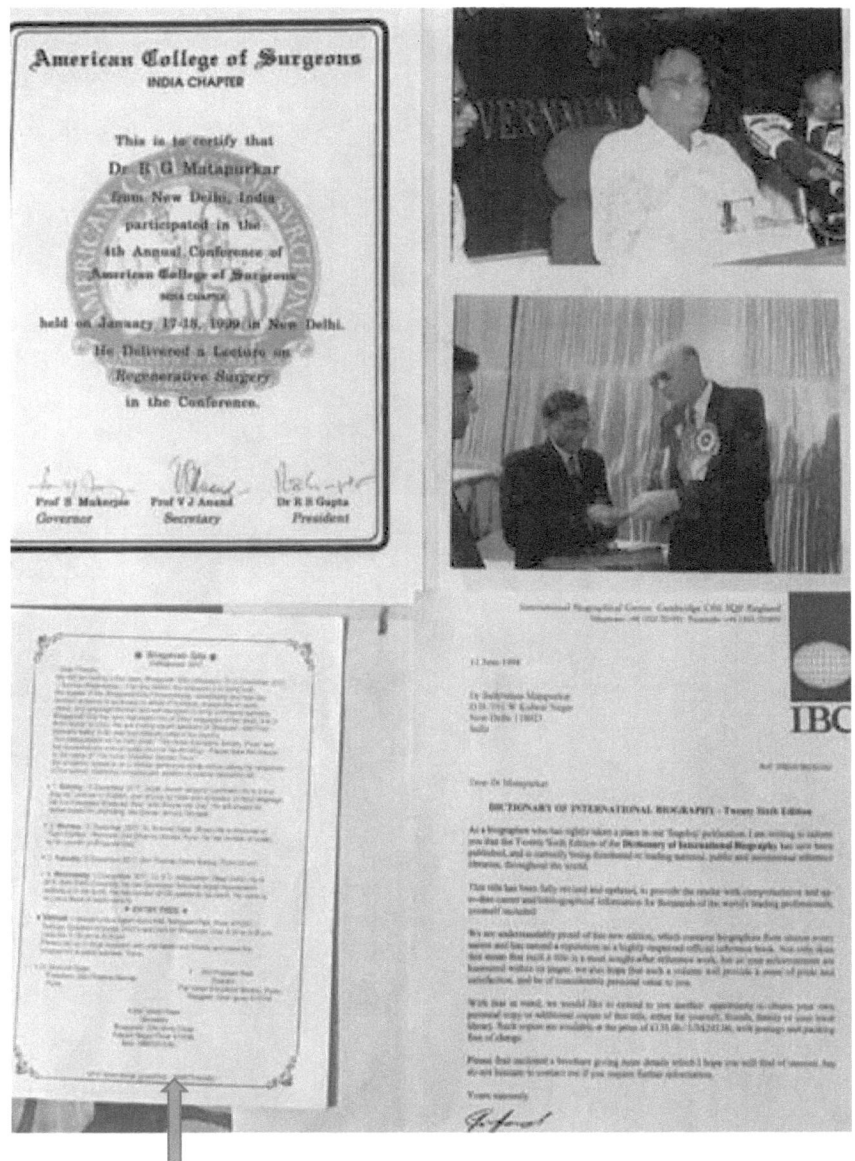

Gita Colloquium at PUNE Maharashtra

INTERNATIONAL PENGUIN PUBLISHING HOUSE

IBC

Hospimedica

THE BOARD OF GOVERNORS

OF

THE NEW YORK ACADEMY
OF SCIENCES

TAKES PLEASURE IN INVITING

B. G. Matapurkar
Safdarjung Hospital
Surgery
New Delhi 110021
INDIA

TO MEMBERSHIP
IN THE ACADEMY

Shri Manohar Joshi, Hon'ble Chief minister MH state

World Environment Day Ministry of environment and Forest
Invited for lecture on raising day of NM of NH, Chaired by Hon'ble
Minister, Shri S. Prabhu.

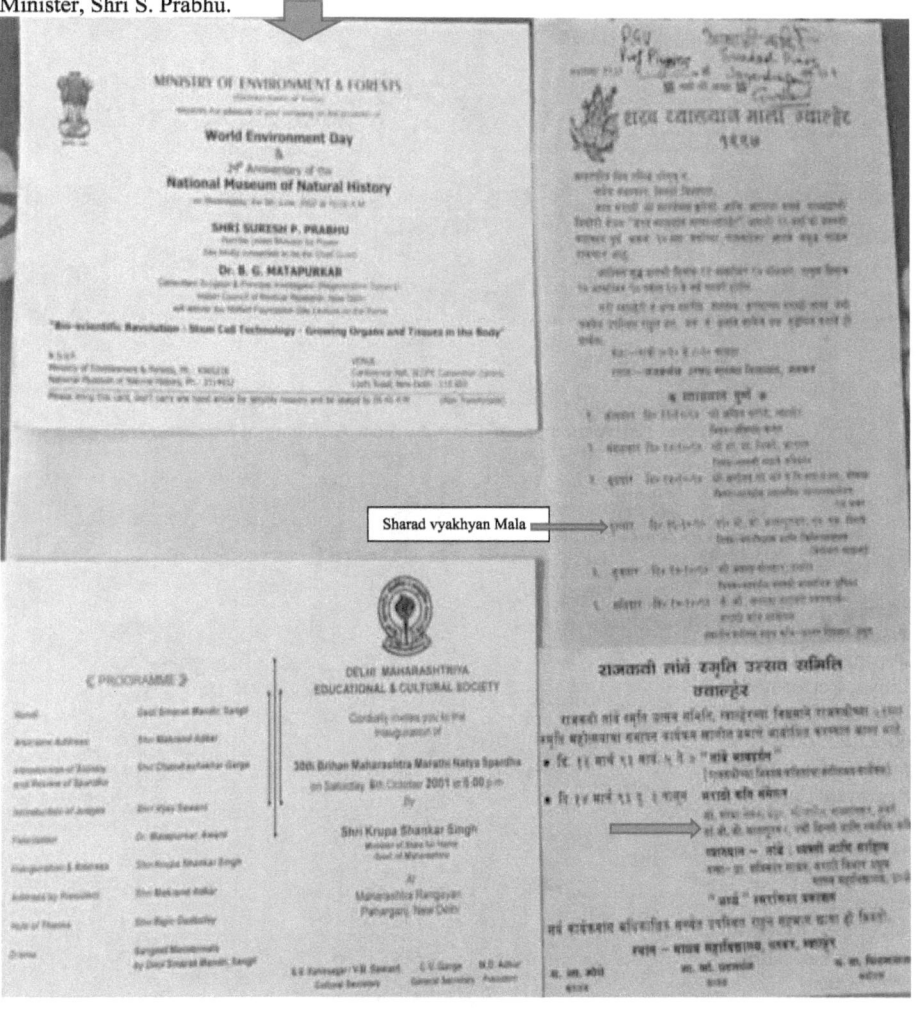

Sharad vyakhyan Mala

world record in Limca book   document from USA.

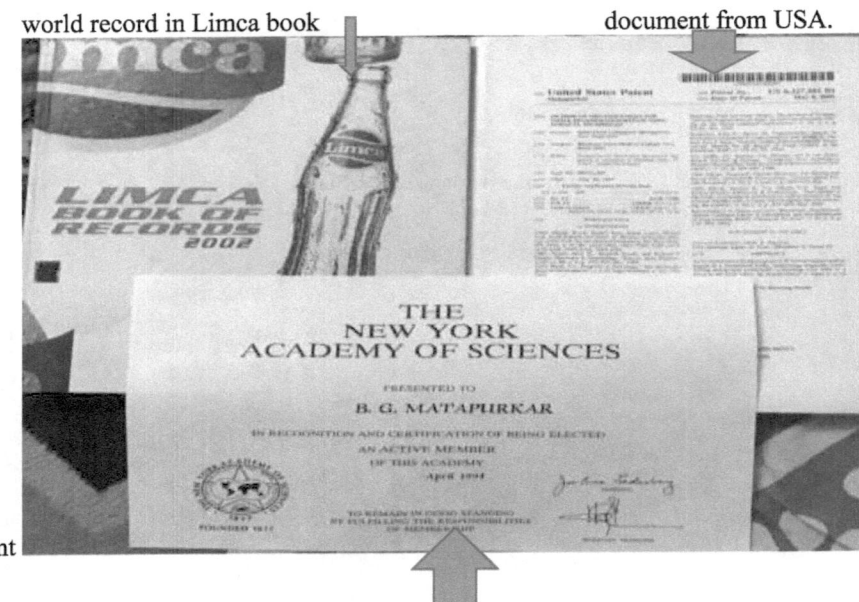

Patent

Elected Active member of The New York Academy of Sciences, USA

When on Deputation to Libya, Performed Rare surgical operation on a patient of Recurrent Ano-rectal fistula successfully, using new technique developed by Dr Matapurkar. This patient was previously operated in UK, France, Germany but failed. Local Arabic news paper Libya

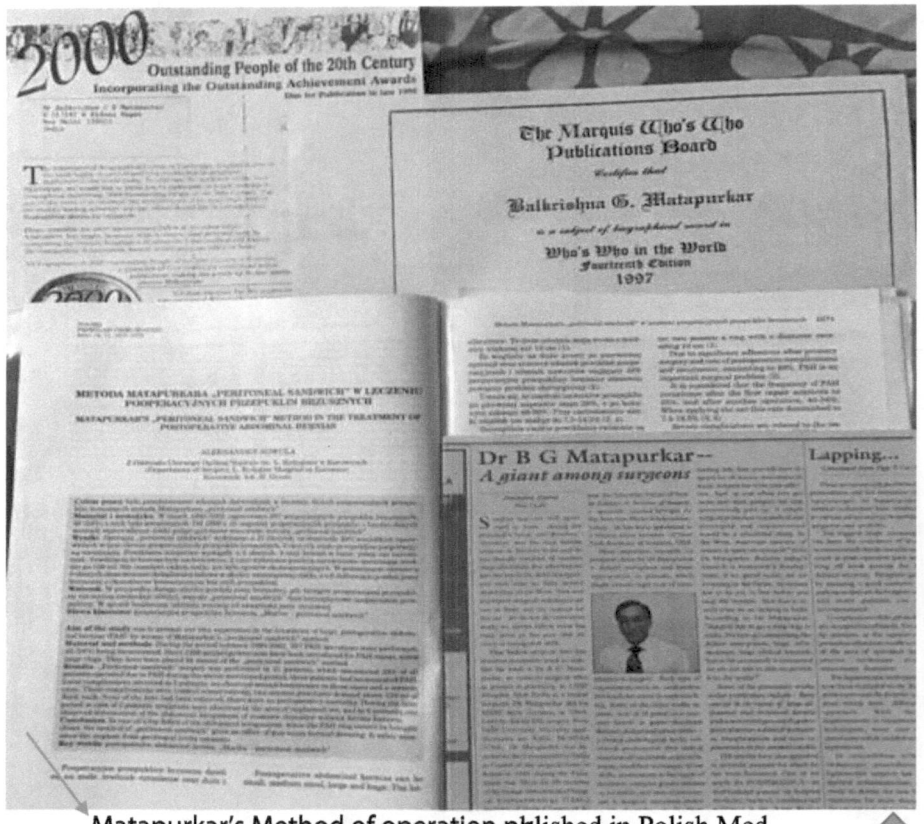

Matapurkar's Method of operation published in Polish Med. Journal and Indian Medical tribune

Biography published in Marquis Whos Who in the world.

REFERENCES

Most of the references are quoted in texts which are at appropriate places.

1. Cosmology Old and New. Prof. G R Jain. Bharatiya DNYANPITH. 2nd Edition. 1991. Page 21.
2. Lost Civilization. India Today. February 11, 2002.
3. THE EARTH. Times Life Books. A. Beiser, 1980, Hongkong.
4. Astrology. Dating MAHABHARATA Two eclipses on 13th day, Dr. S. Balakrishna, 23rd Feb, 2002. Bhishma Parv, Mahabharat.
5. Aryabhatiya by Brahmagupta S. Shukla. New Delhi 1976.
6. Surya Sidhanta, Translation of Ancient Astrological Text. Bapudeva, Varanasi. 1860.
7. Varahamihir's Brihat Sanhita. M Ramkrishna Bhat. Motilal Banarasidas. Publication 1981.
8. Bhagvad – Gita as it is. 2nd Erust, Los Angeles, London, Stockholm, Bombay, Sydney, HongKong.
9. CHHANDOGYOPNISHAD. Gita Press Gorakgpur. Code 582. ISBN. 81-293-0249-7

10. KENOPNISHAD. Swami Chinmayananda. Central Chinmayananda Mission Trust. 2001.

11. Upanishadanche Vidnyanishth Nirupan. Dr. P.V. Vartak. Part 1. Vartak Prakashan 2018.

12. BHAGWAD GITA (Vyas meaning) by Yogi Manohar. Publisher Shri Trimbakrao Manohar Harkare and Shri Dattatray Manohar Harkare,, Tulsi road mahal, Nagpur-440002.

www.ingramcontent.com/pod-product-compliance
Lightning Source LLC
Chambersburg PA
CBHW020722180526
45163CB00001B/76